Essays in Biochemistry

volume 28 1994

73,653

Essays in Biochemistry

Edited by K.F. Tipton

PORTLAND PRESS

Essays in Biochemistry is published by Portland Press Ltd on behalf of the Biochemical Society

Portland Press Ltd
59 Portland Place
London W1N 3AJ
U.K.

British Library Cataloguing-in-Publication Data

Essays in Biochemistry: 28
        I. Tipton, K.F.
        574.19

ISBN 1-85578-016-X
ISSN 0071-1365

Typeset by Portland Press Ltd and printed in Great Britain by Henry Ling Ltd, Dorchester

650

Contents

4 Carnitine and its role in acyl group metabolism
Rona R. Ramsay

5 Folate/vitamin B$_{12}$ inter-relationships
John Scott and Donald Weir

6 Protein kinase inhibitors
Hiroyoshi Hidaka and Ryoji Kobayashi

7 Mitochondrial DNA and disease
Simon R. Hammans

8 PIG-tailed membrane proteins
Anthony J. Turner

9 Horseradish peroxidase: the analyst's friend
Orlaith Ryan, Malcolm R. Smyth and Ciarán Ó Fágáin

10 The renin–angiotensin system
Tadashi Inagami

The Authors

Richard Knowles graduated from the University of Dundee with a B.Sc. in Biochemistry in 1977 and obtained his Ph.D. at St George's Hospital Medical School in the University of London in 1981. He carried out postdoctoral research at the University of Manchester before moving to the Wellcome Research Laboratories in 1985. **Chris Pogson** obtained his B.A. and Ph.D. in Biochemistry at the University of Cambridge. From 1966 until 1984, he held academic positions at the Universities of Bristol, Kent and, latterly, Manchester, where he was Head of the Department of Biochemistry. He was an Editor of the *Biochemical Journal* from 1975 to 1982, and Chairman of the Editorial Board from 1982 to 1987. Currently Head of Biochemical Sciences at the Wellcome Research Laboratories, he also holds visiting professorships in the Universities of Bristol and Kent in the U.K. and in the Universidad de Chile, Santiago, Chile. **Mark Salter** obtained his B.Sc. in Biochemistry from the University of Bristol in 1982 and his Ph.D. at the University of Manchester in 1985. He has been at the Wellcome Research Laboratories since 1985.

Patti A. Quant is the St John's Meres Senior Scholar (Medical Research) in the University of Cambridge where she heads a research group in the Department of Biochemistry involved in application of metabolic control analysis to fatty acid oxidation and ketogenesis. She is also a part-time lecturer and tutor-counsellor for the Open University. Her scientific career began in 1984 when she graduated from the Open University and registered for a Ph.D. in the Department of Biochemistry, University of Cambridge, where she was awarded the Crowther Scholarship (Open University), a Wolfson College (Cambridge) Medical Bursary and the Second Krebs Memorial Scholarship (Biochemical Society). Publications arising from her Ph.D. led to an invited post-doctoral period at the C.N.R.S. in Meudon-Bellevue (Paris) and subsequent election to her current position in 1990.

Ajith Goonetilleke, MBBS MRCP, qualified from St Thomas's Hospital Medical School, London, U.K. He is a Motor Neurone Disease Association Research Fellow and an Honorary Registrar at Charing Cross Hospital, London, U.K. **Jacqueline de Belleroche**, Ph.D., obtained her degree at Trinity College, Dublin and Ph.D. in Neurochemistry at Imperial College, University of London, U.K. She was awarded a Mental Health Foundation Fellowship at the School of Pharmacy and took up a Lectureship in Biochemistry at Charing Cross and Westminster Medical School in 1980, where she is now Reader in Neurochemistry. Her interests lie in applying a range of techniques to understanding neurological and psychiatric disorders. **Roberto J. Guiloff**, MD FRCP, is Consultant Neurologist at Chelsea and Westminster and Charing Cross

Hospitals, London, U.K. He qualified from the University of Chile, Santiago, Chile, and trained in neurology at the National Hospitals for Neurology and Neurosurgery, and King's College Hospital, London, U.K. His research interests include motor neurone diseases and the neurology of HIV infection.

Hiroyoshi Hidaka is Professor and Chairman of the Department of Pharmacology, Nagoya University, Nagoya, Japan. He received his M.D. in 1964 and his Ph.D. in 1968 from Nagoya University. His laboratory research is focused on the pharmacology of protein kinase inhibitors and calmodulin antagonists described in this essay.

Ryoji Kobayashi is associate professor of Nagoya University School of Medicine. He received his M.D. in 1972 and his Ph.D. in 1979 from Shinshu University. He was appointed to his current position in 1989.

Orlaith Ryan graduated in 1990 from University College Galway with an honours B.Sc. in Biochemistry. She undertook research on horseradish peroxidase in the School of Biological Sciences at Dublin City University and obtained her M.Sc. in 1992. She is currently working in industry in Ireland. **Malcolm R. Smyth** studied Biochemistry at Queen's University Belfast and obtained his Ph.D. in Analytical Chemistry from University College London. He acted as Head of the School of Chemical Sciences at Dublin City University from 1990 to 1993, and in 1990 he received a D.Sc. from Queen's University Belfast. He was appointed Professor of Chemistry at Dublin City University in 1992. His main research interests are in the fields of electroanalysis and chromatography, especially in the determination of biologically important substances and in biosensors. He has over 100 publications to his credit and serves on the editorial advisory boards of a number of journals. **Ciarán Ó Fágáin** obtained his B.A. and Ph.D. degrees in Biochemistry at Trinity College Dublin. He then spent 6 years in industry in the production of waste treatment micro-organisms and later in immunodiagnostics development. Since his return to academic life in 1987 he has specialized in protein stability and chemical stabilization. In 1990 he was appointed a lecturer in the School of Biological Sciences at Dublin City University.

Rona R. Ramsay is a research biochemist at UCSF. She studied biochemistry at Edinburgh University and at the University of Cambridge, where she identified the carnitine transport system in 1974. After a postdoctoral period as a Beit Fellow and a Research Fellow of Girton College, Cambridge, she moved to her current laboratory in San Francisco. Her research uses kinetic and protein chemistry techniques to explore the structure and function of enzymes in the pathways of energy generation.

Professor John Scott obtained his B.Sc. from University College Dublin, followed by a Ph.D. from Trinity College Dublin. He then spent 2 years in the University of California at Berkeley before he returned to take up an academic post in the Biochemistry Department of Trinity College, which he still holds.

In the intervening period, he has spent two sabbaticals in the U.S.A., one in Columbia University New York and the other in the University of Minnesota. He currently holds a Personal Chair in Biochemistry. **Professor Donald Weir** qualified in Trinity College, Dublin where he also did an MD on disordered vitamin B_{12} metabolism. As a postgraduate he trained for 4 years in Belfast, London and as a Wellcome Trust Fellow in Edinburgh. He subsequently returned to Trinity College as a lecturer in Clinical Medicine. He is currently the Regius Professor of Physics and Head of the Department of Clinical Medicine in Trinity College.

S.R. Hammans studied medicine at Cambridge University and St Thomas's Hospital, London. He received his M.D. for studies of mitochondrial DNA in mitochondrial myopathies at the Institute of Neurology, Queen Square, London, with Professor A.E. Harding. His research interests are in the genetics of neuromuscular disease. he has recently been appointed as Consultant Neurologist at the Wessex Neurological Centre, Southampton and St Richard's Hospital, Chichester.

Tony Turner graduated in 1969 in Natural Sciences at the University of Cambridge and remained in the Department of Biochemistry to undertake postgraduate research with Keith Tipton. After a year as Royal Society European Research fellow at the Mario Negri Pharmacology Institute in Milan, he returned in 1973 to a lectureship in the Department of Biochemistry, University of Leeds. He is currently Professor of Biochemistry at Leeds and has served as Chairman of the Editorial Board of the Biochemical Journal since 1987. Current research is concerned with neurotransmitter biochemistry and with the structure and physiological functions of cell-surface peptidases which led serendipitously to an interest in GPI-anchors (or PIG-tailed) proteins, the subject of the current essay.

Tadashi Inagami obtained his B.Sc. in 1953 from Kyoto University, Kyoto, Japan, and his M.S. and Ph.D. in biophysical chemistry in 1955 and 1958, respectively, from Yale University. He has been on the faculty of the Department of Biochemistry of Vanderbilt University School of Medicine since 1966. He has been Professor of Biochemistry since 1975, with a joint appointment in Medicine, he is the Director of the Hypertension Research Center. His work has been on biochemical, cell and molecular biological studies of renin, angiotensin receptors, atrial natriuretic factor and its receptors, endothelin, joining peptides, and endogenous inhibitors of Na/K-ATPase.

Abbreviations

ABTS	2,2-azino-di-[3-ethyl benzothiazoline-sulphonate]
ACE	angiotensin I converting enzyme
ALS	amyotrophic lateral sclerosis
AMPA	α-amino-3-hydroxy-5-methyl-4-isoxazole
ANF	atrial natriuretic factor
Ang	angiotensin
AOA	amino-oxyacetate
AP5	2-amino-5-phosphovalerate
APP	amyloid precursor protein
AT1	Ang II receptor subtype 1
BDNF	brain-derived neurotrophic factor
BMAA	β-methylamino-L-alanine
BOAA	β-N-oxalyamino-L-alanine
c.a.t.	computerized axial tomography
$[Ca^{2+}]$	cytosolic free Ca^{2+} concentration
CaM	calmodulin
CAT	carnitine acetyltransferase
c.d.	circular dichroism
Cg	chromogranin
ChAT	choline acetyltransferase
CKI	casein kinase inhibitor
CNTF	ciliary neurotrophic factor
COT	carnitine octanoyltransferase
COX	cytochrome-c oxidase
4-CN	4-chloro-1-naphthol
CPT	carnitine palmitoyltransferase
CSF	cerebrospinal fluid
d.s.c.	differential scanning calorimetry
DA	dopamine
DAB	3,3'-diaminobenzine
DAN	diazoacetyl-D,L-norleucine methyl ester
DMSO	dimethyl sulphoxide
DOPA	3-hydroxy-L-tyrosine
e.i.a.	enzyme immunoassay
e.l.i.s.a.	enzyme-linked immunosorbent assay

e.s.r.	electron spin resonance
EAGMD	experimental autoimmune grey matter disease
EAMND	experimental autoimmune motor neurone disease
EPNP	1,2-epoxy-3-*p*-nitrophenoxypropane
EPSP	excitatory post-synaptic potential
ET	endothelin
FALS	familial amyotrophic lateral sclerosis
G-protein	guanine-nucleotide-binding protein
GABA	γ-aminobutyric acid
GAP	growth-associated protein
GlcNAc	*N*-acetylglucosamine
GPI	glycosylphosphatidylinositol
H-strand	heavy strand
HMG-CoA synthase	3-hydroxy-3-methylglutaryl-coenzyme A synthase
HRP	horseradish peroxidase
5-HT	serotonin
IGF 1	insulin-like growth factor 1
JG	juxtaglomerular
KSS	Kearns-Sayre syndrome
L-strand	light strand
LDL	low-density lipoprotein
LMN	lower motor neurone
LTD	long-term depression
LTP	long-term potentiation
m.r.i.	magnetic resonance imaging
MBTH	3-methyl-2-benzothiazoline hydrazone hydrochloride
MELAS	mitochondrial myopathy, encephalopathy, lactic acidosis and stroke-like episodes
MERRF	myoclonic epilepsy and ragged red fibres
mGT	modified Gomori trichrome stain
MLC	myosin light chain
MM	mitochondrial myopathy
MN	motor neurone
MPP$^+$	1-methyl-4-pyridinium ion
MPTP	1-methyl-4-phenyl-1,2,3,6-tetrahydropyridine

mtDNA	mitochondrial DNA
n.m.r.	nuclear magnetic resonance
N-CAM	neural cell-adhesion molecule
NFT	neurofibrillary tangle
NGF	nerve growth factor
NMDA	N-methyl-D-aspartate
OPD	o-phenylene diamine
PBQ	p-benzoquinone
PDPK	proline-directed protein kinase
PEI	polyethylenimine
PEO	progressive external ophthalmoplegia
pEPSP	population excitatory post-synaptic potential
PHF	paired helical filament
PI	phosphatidylinositol
PI-PLC	phosphatidylinositol-specific phospholipase C
PIG	phosphatidylinositol-glycan
PMA	progressive muscular atrophy
PNH	paroxysmal nocturnal haemoglobinuria
RAS	renin–angiotensin system
RFLP	restriction fragment length polymorphism
RRF	ragged red fibres
SAH	S-adenosylhomocysteine
SAM	S-adenosylmethionine
SCD	sub-acute combined degeneration
Sg	secretogranin
SHR	spontaneously hypertensive rat
SOD	superoxide dismutase
SPDP	N-succinimidyl 3-(2-pyridylthio) propionate
SSCP	single-strand conformational polymorphism
TCNQ	tetrathiofulvalinium tetracyano-quinodimethanide
TDO	tryptophan 2,3-dioxygenase
TH	tyrosine hydroxylase
TMB	3,3′,5,5′-tetramethylbenzidine
TPA	12-O-tetradecanoylphorbol-13-acetate
TRH	thyrotrophin-releasing hormone
TTF	tetrathiofulvalene

UMN	upper motor neurone
VSG	variant surface glycoprotein
WP-ALS	Western Pacific amyotrophic lateral sclerosis

Metabolic control

Mark Salter, Richard G. Knowles and Chris I. Pogson

Biochemical Sciences, Wellcome Research Laboratories, Langley Court, Beckenham, Kent, BR3 3BS, U.K.

Introduction

What do we mean by the word 'control'? In normal use, the words 'control' and 'regulation' have overlapping meanings, but in discussing metabolism and its regulation this overlap can (and does) lead to confusion. We therefore need more precise meanings for these words.

Regulation is the overall process whereby a pathway is regulated. For each step in the pathway this is expressed in terms of two properties: (*i*) **regulability**, reflecting the extent to which the rate of the step can be increased or decreased, and (*ii*) **control**, reflecting the extent to which a given change in the rate of the step affects the pathway.

The concept of regulability of an enzymic reaction, by allosteric effectors, by post-translational modification (such as phosphorylation), or by induction or repression of enzyme synthesis, is quite a simple and well-understood one. The concept of control is much less easy to grasp, and it is this important property, its quantification and practical importance that form the main concern of this article.

Early students of metabolism concerned themselves with the sequences of intermediates in pathways and with the properties of individual enzymes. Much attention was given to changes in the activity of isolated enzymes after various treatments *in vivo* or *in vitro* (i.e. regulability), but little time was spent in considering how these changes in enzyme activity might be reflected in the overall function of the pathway.

More recently, many groups have tried to describe the degree of control exerted by enzymes on the rate of (i.e. flux through) specific pathways. Enzymes have been described as rate limiting, even as being responsible for 'the slowest step'. The first term is, however, an over-simplification quantitatively — the second term is usually an impossibility!

Consider the hypothetical pathway below where A, B, C and D are intermediates (metabolites) and E_1, E_2 and E_3 are the enzymes responsible for their

$$
A \underset{90}{\overset{100}{\rightleftharpoons}} B \underset{0.1}{\overset{10.1}{\rightleftharpoons}} C \underset{10}{\overset{20}{\rightleftharpoons}} D \quad (1)
$$

$$
\quad E_1 \qquad\qquad E_2 \qquad\qquad E_3
$$

interconversion in the pathway; these enzymes may be at varying degrees of saturation with their respective substrates A–D. The numbers represent the rates of 'forward' and 'backward' reactions catalysed by the enzymes. We can obtain a value for the flux through the pathway (i.e. the net rate at which substrate A is converted to product D) by subtracting the 'backward' from the 'forward' rate; in this example, the value is 10. When a pathway is at steady state — that is, the rates of the 'forward' and 'backward' reactions are constant with time — there can be no such thing as a 'slowest step', because all enzymes are catalysing reactions with the same net rate (flux). We cannot even assume that the enzyme catalysing the slowest unidirectional (i.e. either 'forward' or 'backward') reaction is the major controlling enzyme. Often in past literature, and still occasionally today, we can read of reactions that are described as being catalysed by 'the rate-limiting enzyme', without there being any real evidence as to whether these reactions are sites of appreciable control or not. In fact, every enzyme in a pathway is rate limiting to some degree, in the sense that it exerts some influence, however small, on the overall flux; it follows that the rate limitation, or control, is shared (but not necessarily, or even usually, equally) between all of the enzymes in the pathway. Control may reside mainly with one enzyme, but it does not have to do so. If conditions in the pathway change (e.g. there is a change in the activity of one of the enzymes), then the pathway will adjust itself to a new steady state and the distribution of control between the enzymes will also change. It is important to recognize, therefore, that the distribution of control is a global property of the whole system and not an intrinsic enzymic property.

Clearly we should go further than merely qualitative descriptions of control in pathways and try to describe systems in numerical terms. One way of achieving this is to make use of what is now known as 'Metabolic Control Theory', based on concepts originated by Higgins, Kacser and Burns, and Heinrich and Rapoport. This theory and its applications to real experimental systems will be described in the following sections.

General description and definition of terms

Consider the hypothetical pathway below where A–D are the metabolites and E_1–E_3 are the enzymes. If the activity of one of the enzymes in the pathway

$$A \xrightleftharpoons[]{E_1} B \xrightleftharpoons[]{E_2} C \xrightleftharpoons[]{E_3} D \qquad (2)$$

(e.g. E_1) is changed, we can measure the effect that this change has on the flux J. We can find out how much the net enzyme activity (i.e. the 'forward' minus the 'backward' rate, represented by v) of E_1 has changed, by determining it before and after the change. This net activity is measured under conditions where the concentrations of substrates, products and other effectors of E_1 occurring in the pathway (before the change was made), are held constant. We can repeat this process with larger or smaller changes in the activity of E_1 and so generate a curve relating activity (v) to flux (J). We now have the information needed for determining the degree of control exerted by E_1 on the pathway. This is expressed as the **flux control coefficient** (C^J), the fractional change in flux ($\partial J/J$) resulting from a fractional change in net enzyme activity ($\partial v/v$), as shown in eqn (3). The symbol ∂ is used to represent infinitesimal changes, where the fractional changes are very small (i.e. tend to zero).

$$C^J_{E_1} = \frac{\partial J / J}{\partial v / v} \qquad (3)$$

The greater the value of C^J, the greater the control exerted by that enzyme on the pathway. The superscript refers to the parameter controlled (in our example, J stands for pathway flux); the subscript refers to the parameter being altered (here the enzyme E_1).

There are a number of simple properties of flux control coefficients:

- The **Summation Theorem** states that the sum of all the flux control coefficients in a pathway is always 1.0; this means that, for most cases, the value of any single coefficient can vary from nearly zero to nearly 1.0. In branched pathways the values can even be negative, or more than 1.0.

- There are as many flux control coefficients as there are steps in a pathway, which may include transport processes and non-catalysed, as well as enzymic, reactions. Individual coefficients for a part of a metabolic sequence can, however, for various reasons, be grouped together to give 'global' coefficients (which must still add up to 1.0 for the whole pathway).

- The more similar changes in the overall flux are to changes in the activity of a particular enzyme, the closer is the flux control coefficient for the step catalysed by that enzyme to 1.0.

We can also, in an analogous way, describe the control an enzyme has on the concentration of a particular metabolite rather than on the flux. This effect is described quantitatively by the **concentration** or **metabolite control coefficient**; compare eqn (3) with eqn (4), where S is the concentration of the metabolite.

$$C_{E_1}^S = \frac{\partial S / S}{\partial v / v} \qquad (4)$$

This coefficient has the following properties: (i) there are as many metabolite control coefficients as there are combinations of enzymes and metabolites in a pathway; (ii) the sum of all metabolite control coefficients for any given metabolite is zero; and (iii) the values of metabolite control coefficients vary with conditions (as do flux control coefficients) and can be >1 or <1.

There are other coefficients that help to characterize the pathway control structure.

The **elasticity coefficient** (ε) describes the effect of changes in the concentration of a substrate, or other effector, of an enzyme on the net rate (i.e. v) of that enzyme. Again compare eqn (5) with eqns (3) and (4).

$$\varepsilon_S^{E_1} = \frac{\partial v / v}{\partial S / S} \qquad (5)$$

Elasticity coefficients can be positive or negative, but, unlike flux or metabolite control coefficients, do not give us direct information about the control exerted by an enzyme. When we combine elasticity coefficients in certain ways, however, we can calculate control coefficients. For example, look again at our hypothetical pathway (2), and assume we have data which allows us to calculate elasticity coefficients for metabolite B on enzymes E_1 and E_2. The ratio of these elasticity coefficients is related to the flux control coefficients of the two enzymes as shown in eqn (6).

$$\frac{\mathcal{E}_B^{E_1}}{\mathcal{E}_B^{E_2}} = -\frac{C_{E_2}^J}{C_{E_1}^J} \tag{6}$$

If, independently, we know, or can determine, the flux control coefficient of either E_1 or E_2, then we can easily calculate the coefficient for the remaining enzyme.

We can also calculate a control coefficient from the relationship between an external elasticity coefficient and a **response coefficient**. An external elasticity coefficient describes the effect of an externally added effector Q (which can be a substrate of just one enzyme, an activator or an inhibitor) on the net rate catalysed by enzyme E, as shown in eqn (7).

$$\mathcal{E}_Q^E = \frac{\partial v / v}{\partial Q / Q} \tag{7}$$

Response coefficients (R) are of two kinds: one describes the effect of an effector Q on pathway flux [eqn (8)], and the other the effect of Q on the concentration of a metabolite [eqn (9)]. Note that response coefficients are, like control coefficients (but unlike elasticity coefficients), global properties of the pathway:

$$R_Q^J = \frac{\partial J / J}{\partial Q / Q} \tag{8} \qquad\qquad R_Q^S = \frac{\partial S / S}{\partial Q / Q} \tag{9}$$

The response, elasticity and control coefficients are simply related as shown in eqns (10) and (11).

$$C_E^J = \frac{R_Q^J}{\mathcal{E}_Q^E} \tag{10} \qquad\qquad C_E^S = \frac{R_Q^S}{\mathcal{E}_Q^E} \tag{11}$$

The next section shows how the theory can be put into practice with real biological systems.

Practical determination of control coefficients

In any particular study it is a good idea to determine control coefficient(s) in more than one way; we can thus check on any assumptions made when just one method is employed. This does not, however, provide too many problems

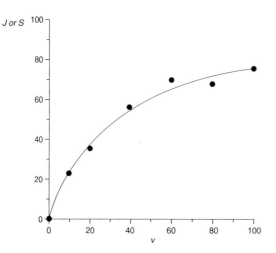

Figure 1. Effect of changing enzyme (v) on metabolic flux through the pathway (J) or on a metabolite in the pathway (S)

because control coefficients can be obtained in several ways; three examples are given below.

One method involves changing the activity of the enzyme of interest — either by changing its K_m (for example, by phosphorylation) or its $V_{max.}$ (for example, by induction or other manipulation at the gene level) — and measuring the related changes in flux (for flux control coefficients) or in the concentration of a chosen metabolite (for metabolite control coefficients). The result of doing this is shown in Figure 1. The control coefficient for a given value of v can be calculated, using eqns (3) or (4), from the tangent to the curve at that point. In this example, as v increases, the change in J (flux) or S (metabolite concentration) becomes less responsive, reflected in progressive decreases in the respective control coefficients.

The second method involves the use of an external effector which may be a substrate or selective inhibitor. The effects of varying the concentration of the effector (Q), both on the net activity (v) of the chosen enzyme (E) in the pathway (under 'pathway conditions') and on overall flux (J), or on the concentration of an intermediate (S), are measured [see eqns (8–11)]. Typical data are shown in Figure 2, where Q is an inhibitor of enzyme E.

Control coefficients can also be obtained from the elasticity coefficients where there are two adjacent enzymes with a common intermediate. Consider the pathway segment shown below:

$$A \xrightleftharpoons{E_1} B \xrightleftharpoons{E_2} C \tag{12}$$

We need first to calculate the elasticities of E_1 and E_2 for metabolite B. If we know the equilibrium constants for these enzymes and can measure the concentrations of their ligands A, B and C, then we can use eqns (13) and (14): where v_f and v_r are the 'forward' and 'backward' (or 'reverse') rates, V is the

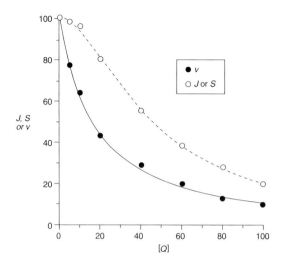

Figure 2. Effect of the concentration of an added effector (Q; here an inhibitor) on enzyme activity (v), metabolic flux (J) or the concentration of a metabolite (S)

relavent maximum rate, $K_{eq.}$ is the equilibrium constant and is the **Mass Action Ratio** (the ratio of the concentrations of products to the concentrations of substrates actually found in the pathway) for E_1 (eqn 13) or E_2 (eqn 14). Note that the numerical value of the Mass Action Ratio approaches that of the equilibrium constant as the enzymic reaction approaches equilibrium.

$$\varepsilon_B^{E_1} = \frac{1}{1 - \Gamma / K_{eq.}} - \frac{v_f}{V_f} \quad (13) \qquad \varepsilon_B^{E_2} = \frac{-\Gamma / K_{eq.}}{1 - \Gamma / K_{eq.}} - \frac{v_r}{V_r} \quad (14)$$

When we have calculated the two elasticity coefficients ε_B^{E1} and ε_B^{E2} we can use eqn (6) to obtain the ratios of the control coefficients. If one of these is already known, then the other is simply obtained. If the pathway consists only of our two enzymes E_1 and E_2, then we can use our knowledge that the sum of all flux control coefficients is 1.0 (i.e. $C_{E1}^J + C_{E2}^J = 1.0$) and of metabolite control coefficients zero (i.e. $C_{E1}^S + C_{E2}^S = 0$).

Solution of eqns (13) and (14) is often not so hard as it looks, because, in many instances, the rates of the enzymes involved are well below maximal; this means that we can make the simplifying assumption that the value of the second term (v/V) is very small and can be disregarded.

When a pathway consists of many steps or where it is difficult to obtain specific data, we can make life easier by combining 'adjacent' enzymes into groups and considering each group as if it were one enzyme alone — that is, we can determine flux control coefficients for portions of pathways, remembering that the Summation Theorem still applies. As an example, we might aggregate all cytosolic reactions using ATP (whether membrane pumps or kinases), into one 'ATPase' whose influence on mitochondrial respiration could be measured.

The following section shows how these methods have been exploited to determine the control structure of a number of pathways.

Some practical examples

Metabolic Control Theory has been applied to a diversity of biological systems, some relatively simple, others more complex. In this section, we show how theory can be applied to practical examples, and discuss the conclusions which can be made from these studies.

The major route of tryptophan metabolism in mammals is through the kynurenine pathway in the liver (Figure 3). Tryptophan 2,3-dioxygenase (TDO) was the first mammalian enzyme whose activity was found to increase after steroid treatment. It was generally regarded (as we now recognize, erroneously) as the 'rate-limiting' enzyme of the pathway. The flux control coefficient of TDO was determined by treating rats in various ways that change the amount, and hence the activity, of the enzyme. Liver cells, isolated from the treated rats, were incubated with tryptophan and the oxidative flux through TDO was measured (Figure 4).

Note that, as TDO activity increases, the rate of increase of flux progressively diminishes. This is shown in another way in Figure 5. The flux control coefficient of TDO decreases as the amount of enzyme increases. Under basal conditions (when $v = 12$) C_{TDO}^{J} is 0.75, but after maximal induction ($v = 110$) this value falls to only 0.25, far from justifying the term 'rate-limiting'.

Passage across the plasma membrane is a step not always included in textbook diagrams of metabolic pathways, but, as we shall see, it can exert control over the pathway. The response coefficient is easily determined from a plot of flux against the concentration of tryptophan added to the medium (eqn 8). The elasticity coefficient is obtained from direct measurements of the rate of trans-

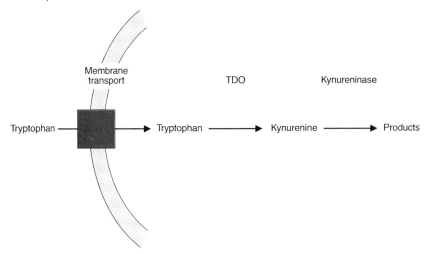

Figure 3. Kynurenine pathway of tryptophan metabolism in liver

port as a function of added tryptophan (eqn 7). The flux control coefficient for membrane transport (C_{TR}^J) is obtained by dividing the response coefficient by the elasticity coefficient, as shown in eqn (10). Figure 5 shows how C_{TR}^J changes as the activity of TDO changes, from 0.25 at basal TDO activities to 0.75 in cells with maximal TDO.

Figure 5 illustrates two important points.

- Although the transport step itself is neither induced nor repressed under conditions where TDO activity changes, its role in control alters significantly, and inversely with that of TDO. Thus changes at one step will affect the control characteristics of others.

- The sum of the two control coefficients (C_{TDO}^J and C_{TR}^J) equals 1.0. Because the sum of all flux control coefficients, by definition, equals 1.0, control of the pathway is, within experimental error, shared between these two steps. To all intents and purposes, therefore, reactions subsequent to the TDO step exert negligible control.

One of the reactions further down the pathway is catalysed by kynureninase (Figure 3), which requires pyridoxal 5-phosphate for activity. Aminooxyacetate (AOA) reacts covalently with pyridoxal 5-phosphate and can be used to titrate kynureninase activity (Figure 6). When incubated with 'normal' liver cells, AOA has only a small effect on flux (curve A), despite its relative potency against the isolated enzyme (curve C); this is what we would expect for a step with a very low flux control coefficient. Kynureninase activity is substantially lowered in pyridoxine deficiency, and, under these conditions, AOA proves to be a much better inhibitor of flux (curve B). Again the message is that the disposition of control, the effectiveness of inhibitors against enzymes in a pathway, and the individual flux control coefficients, can and do vary from one circumstance to another.

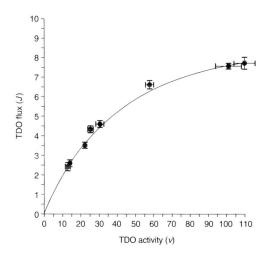

Figure 4. Effect of varying TDO activity on the catabolism of tryptophan in intact cells, expressed as TDO flux
Each point represents results from cells obtained from rats treated in different ways so as to alter the amount of TDO protein (and hence activity). Compare with Figure 1.

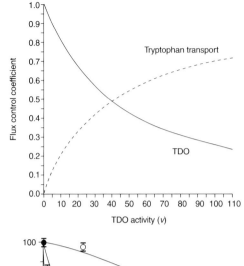

Figure 5. Changes in the flux control coefficients for TDO and transport of tryptophan across the plasma membrane as a function of changes in total TDO activity

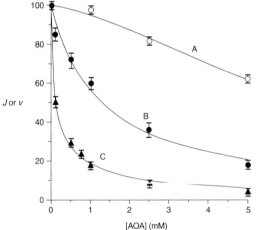

Figure 6. Titration of kynureninase with AOA

Curves A and C show the effect on metabolic flux and isolated enzyme activity, respectively; Curve B shows the effect on metabolic flux in cells from a pyridoxine – deficient rat.

Groen and his colleagues have applied the Theory to the control of mitochondrial respiration. They used different inhibitors to quantify the distribution of control for respiration between the several components in different states of respiration. The results (in Figure 7) show beautifully how control can vary with changes in pathway conditions (i.e. states of respiration) and exemplify the problem and lack of logic in searching for 'the rate-limiting reaction'.

Why bother?

The development of the Metabolic Control Theory now allows us to talk about the control structure of a pathway in quantitative terms. But does this have any impact in the real world or is it merely of theoretical interest? The techniques of molecular biology allow us to alter the amounts or the kinetic properties of specific enzymes in a given system. Such changes are, however, not easy to bring about, and it is obviously to our advantage if we can predict the outcomes in advance. Metabolic Control Theory permits us to do this. For

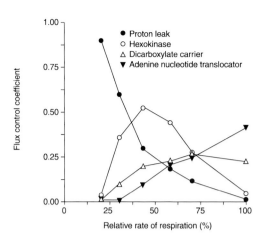

Figure 7. Effects of changes in the rate of respiration on the flux control coefficients of four components in a mitochondrial incubation.

example, if we understand the control of a fermentation process, we can alter those enzymes with greatest control coefficients in such a way as to maximize flux to end-products or to increase the concentrations of specific intermediary metabolites. Changes to enzymes with low control coefficients will probably have less far-reaching effects. A second example may be found in the drug-discovery process. If we are looking for compounds effective against specific pathways or processes *in vivo*, then a good strategy is to look for inhibitors of the enzymes with the highest flux control coefficients. As a guide, we can divide the K_i value for an inhibitor by the flux control coefficient of the relevant enzyme to get a rough idea of the expected potency *in vivo*. Thus, if we have a K_i of 1 mM, but a value for C_E^J of only 0.02, we can anticipate that the pathway will be inhibited by 50% only at a concentration of inhibitor of approximately 1.0/0.02 = 50 mM; a less potent compound acting on an enzyme of greater control significance may be much more effective *in vivo*. Realization of this fact can help to guide us in identifying good candidate enzymes as targets for therapeutic intervention, and thus to increase the efficiency of drug discovery.

As we expand our knowledge of the biochemistry of complex physiological and pathological processes and wish to modify these for our own benefit, so our need for quantitative treatments grows; Metabolic Control Theory provides a simple, elegant and practical way of fulfilling this need.

Further reading

Fell, D.A. (1992) Metabolic Control Analysis: a survey of its theoretical and experimental development. *Biochem. J.* **286**, 313-330

Groen, A.K., Wanders, R.J.A., Westerhoff, H.V., Van Der Meer, R. & Tager, J.M. (1982) Quantification of the contribution of various steps to the control of mitochondrial respiration. *J.Biol.Chem.* **257**, 2754-2757

Kacser, H. & Porteous, J.W. (1987) Control of metabolism: what do we have to measure? *Trends Biochem. Sci.* **12**, 5-15

Pogson, C.I., Knowles, R.G. & Salter, M. (1989) The control of aromatic amino acid catabolism and its relationship to neurotransmitter amine synthesis. *Crit. Rev. Neurobiol.* **5,** 29-64

Salter, M., Knowles, R.G. & Pogson, C.I. (1986) Quantification of the importance of individual steps in the control of aromatic acid metabolism. *Biochem. J.* **234,** 635-647

The role of mitochondrial HMG-CoA synthase in regulation of ketogenesis

Patti A. Quant

Department of Biochemistry, University of Cambridge, Tennis Court Road, Cambridge CB2 1QW and St. John's College, Cambridge CB2 1TP, U.K.

Introduction: what is ketogenesis?

Ketogenesis is a normal metabolic process that occurs exclusively in the mitochondrial compartment of certain cells (Figure 1) in most species, and produces ketone bodies, acetoacetate and β-hydroxybutyrate from fatty acids. (Acetone is formed by spontaneous decarboxylation of acetoacetate when the latter is present in abnormally high concentrations; however, as it is not metabolized further, but exhaled or excreted, its contribution to ketogenesis is insignificant and will not be considered in this article.) Liver mitochondria are the main producers of ketone bodies, but kidney, small intestine[1] and white adipose tissue mitochondria[18] also synthesize ketone bodies in some metabolic states or during certain stages of development (Figure 2).

Ketone bodies are small soluble 'fat' molecules which are vital oxidizable chemical fuels for the brain and peripheral tissues in some metabolic (prolonged starvation)[2], nutritional (high-fat diet) and developmental (during suckling)[1] states. Unlike fatty acids, ketone bodies can cross the blood-brain barrier and during prolonged starvation, when hepatic glycogen stores are depleted, the brain is able to derive up to 75% of its energy needs from oxidation of ketones. As well as being important metabolic fuels which enable glucose to be

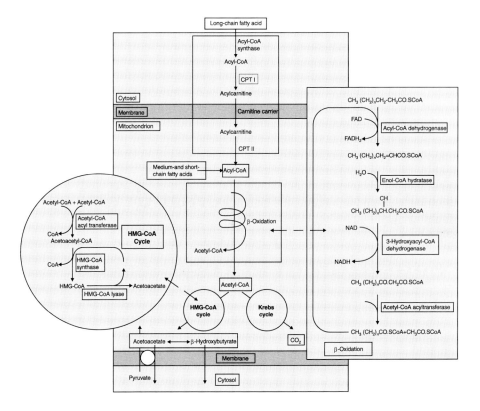

Figure 1. Intracellular pathway of ketogenesis
CPT I, carnitine palmitoyltransferase I; HMG-CoA synthase, 3-hydroxy-3-methylglutaryl-coenzyme A synthase.

spared, ketone bodies are substrates for myelination of the neonatal brain[3] and, to a lesser extent, precursors for lipogenesis in the lactating mammary gland[4] (Figure 2).

The physiological pathway of ketogenesis

Fatty acids, the pathway substrates, are released into the blood from fat stores in adipose tissue and taken up by the liver (or other tissues) for ketone body synthesis, or oxidized directly by muscle to CO_2 (Figure 3).

Long-chain fatty acids cross the plasma membrane of target cells and are activated to acyl-CoAs in the cytoplasm by an ATP-utilizing acyl-CoA synthase. The acyl-CoAs, which cannot traverse mitochondrial membranes without modification, are esterified to acyl-carnitines by carnitine palmitoyltransferase (CPT) I (EC 2. 3. 1. 21) and transported across the inner mitochondrial membrane, in exchange for carnitine, by the carnitine carrier. CPT II catalyses the reverse reaction, converting acyl-carnitines back to acyl-CoAs which are

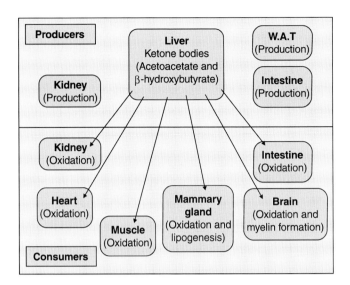

Figure 2. Ketone body producing and consuming tissues
WAT, white adipose tissue.

released into the matrix to undergo β-oxidation. Medium- and short-chain fatty acids can cross mitochondrial membranes independently of CPT I, the carnitine carrier and CPT II, and are activated to the CoA-esters by ATP-utilizing reactions *in situ* before undergoing β-oxidation (Figure 1).

The β-oxidation spiral comprises a series of four enzyme-catalysed reactions (acyl-CoA dehydrogenase, enol-CoA hydratase, 3-hydroxyacyl-CoA dehydrogenase and acetyl-CoA acyltransferase) which cleave a two-carbon acetyl-CoA moiety at each complete cycle. The residual moiety re-enters the spiral for further oxidation until, at the final turn of the cycle, two acetyl-CoA molecules are produced. Two oxidation steps in the sequence release reducing equivalents $FADH_2$ and NADH for re-oxidation in the electron-transport chain. Odd-number carbon chain fatty acids yield propionate, a three-carbon molecule, in the final step.

Acetyl-CoA, the product of β-oxidation from even-chain fatty acids, can be transported out of the mitochondria on the citrate carrier (in the form of citrate) and then used as a substrate for lipogenesis, be further oxidized to CO_2 by the reactions of the Krebs' cycle in the mitochondrial matrix or be converted to ketone bodies, acetoacetate or β-hydroxybutyrate, by the enzymes of the 3-hydroxy-3-methylglutaryl-CoA (HMG-CoA) cycle (acetyl-CoA acyltransferase, HMG-CoA synthase and HMG-CoA lyase) (Figure 1). β-Hydroxybutyrate dehydrogenase interconverts acetoacetate to β-hydroxybutyrate in an NADH/NAD-linked reaction.

Ketone bodies leave the mitochondria by diffusion across the mitochondrial membranes or in exchange for pyruvate (Figure 1).

Therefore, while the physiological pathway of ketogenesis involves more than one compartment of several tissues (Figure 3), the part of the pathway committed to ketogenesis is completely contained within the mitochondria of cells. The network of these reactions involved in ketone body production is

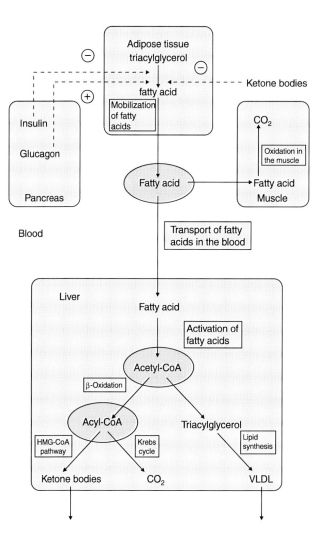

Figure 3. Branched nature of the physiological pathway of ketogenesis
VLDL, very low-density lipids; − signifies inhibition; + signifies activation.

well-established (Figure 1), but the identity and quantitative nature of all the control sites, under different sets of metabolic conditions, has not been fully described. Both the pathways of ketone body production and consumption (Figure 2) probably contribute significantly to control of ketogenesis. (When demand for ketone bodies is high, e.g. during suckling, rapid utilization could increase the rate of synthesis.) The physiological ketogenic pathway is complex (involving several moiety-conserved cycles; Figure 1) and highly branched (Figure 3) with each branchpoint representing a possible site of control. Any model attempting to describe how hepatic ketogenesis is regulated must account for control over both the rate and the direction of flux at each of the branch points, and demonstrate how these controls are achieved.

To date, descriptions have been intuitive and qualitative[13] and a full quantitative metabolic control analysis (Chapter 1) is lacking, although results of preliminary experiments have been published[5,6].

However, it is clear that hepatic ketogenesis is regulated from outside and within the liver as there is often no clear correlation between rate of release of fatty acids, the pathway substrates, from adipose tissue and rate of ketone body production. External controls may include those reactions involved in mobilizing and releasing fatty acids from adipose tissue; the rate of delivery of the fatty acids to the liver; regulation of blood flow; secretion of hormones by the pancreas (Figure 3); nervous inhibition; those reactions producing ketone bodies, (in the kidney, intestine or white adipose tissue), and those in the brain and peripheral tissues that oxidize ketone bodies, or use them as substrates for lipogenesis and myelin formation[7] (Figure 2). Possible intrahepatic control sites include activation of long-chain fatty acids by acyl-CoA synthase; transfer of carbon for lipid synthesis in the cytosol; transport of acyl-CoAs into mitochondria (involving CPT I, carnitine carrier and CPT II); the enzymes of β-oxidation; the HMG-CoA pathway and/or the Krebs' cycle (and, therefore, the enzymes of the electron transport chain and the ATPase which re-oxidize reducing equivalents), and efflux of ketone bodies from the mitochondria via pyruvate exchange[7] (Figure 1).

This article is concerned mainly with regulation of those reactions of the pathway occurring in association with or within the mitochondrial compartment of the liver cell (Figure 1).

The rate of ketogenesis is regulated by hormones, nutrition and development

The physiological rate of ketogenesis varies[4,8]. In adults who are normally fed, it is low, even during periods of stress, when high levels of vasopressin maintain a low rate of ketogenesis despite high levels of non-esterified fatty acids in the circulation. However, it is elevated under conditions where the glucagon:insulin ratio is high, i.e. during certain stages of development, e.g. during the suckling period when the demand for ketone bodies is high[1,9]; with a high-fat diet; during late pregnancy; with lactation; with prolonged starvation when ketones provide an alternative fuel to spare glucose levels[2] and with sustained exercise. Under these conditions the elevated plasma ketone levels (physiological ketosis) are beneficial to the individual and homoeostatic mechanisms ensure (that in humans) they never exceed safe levels (8 mM).

However, with some diseases, e.g. untreated juvenile onset diabetes, where circulating glucagon levels are high and insulin levels are low[10], and episodes of bovine or ovine pregnancy toxaemia, the plasma ketone body concentrations exceed 20 mM (pathological ketoacidosis) and become life-threatening. Because ketone bodies are acids the blood cannot buffer such a high acid load and toxaemia results. Such ketoacidosis can result in coma and death.

These observations suggest that hormones (predominantly peptide hormones such as insulin, glucagon and vasopressin, but also thyroid hormones); nutrition (the percentage of fat in the diet) and development are effectors of ketogenesis[1-3,7-14].

The role of malonyl-CoA inhibition of CPT I

The activity of CPT I, the enzyme, with the carnitine carrier and CPT II, responsible for transport of long-chain fatty acids into the mitochondria of cells (Figure 1), is controlled by malonyl-CoA inhibition[15]. In turn, the synthesis of malonyl-CoA is enhanced by the stimulatory action of insulin and depressed by the inhibitory action of glucagon on acetyl-CoA carboxylase, the first enzyme in the cytosol committed to lipogenesis. This avoids futile cycling (simultaneous oxidation and synthesis) of fatty acids. Since this mechanism was proposed, most studies of liver CPT I have focused on this control mechanism which is generally assumed to be important in regulation of ketogenic flux in livers of normal fed adults[15]. However, the capacity of the enzyme and its effector to control ketogenesis *in vivo* has not been quantitatively measured[13]. The popular view that CPT I is the 'rate-limiting' enzyme of fatty acid oxidation and ketogenesis is too simple as the model is not consistent with rapid changes in the rate of ketogenic flux observed during certain transition periods when CPT I activity does not alter over the same timescale: the fetal-neonatal and suckling-weaning transitions; reversal of diabetic ketoacidosis with insulin-treatment; the starved-to-fed transition; and during periods of high stress[1,2,4-14]. Furthermore, during the suckling period in rats, 40% of the fatty acids derived from milk are medium- and short-chain and enter the mitochondria independently of CPT I, suggesting significant control over the pathway flux is exerted at other step(s).

There is accumulating indirect evidence that significant shorter-term controls over ketogenesis may be invested at CPT I[16] and other (intramitochondrial) sites.

Strong evidence has been presented that in addition to CPT I, controls of mitochondrial HMG-CoA synthase (EC 4.1.3.5), the second enzyme of the HMG-CoA pathway (Figure 1), may contribute to regulation of ketogenic flux[1,5,7,9,11-14,17,18].

The role of succinylation of mitochondrial HMG-CoA synthase

Mitochondrial HMG-CoA synthase catalyses the conversion of two substrates, acetyl-CoA and acetoacetyl-CoA, to HMG-CoA in a 'Bi-Bi-Ping-Pong' type reaction (Figure 4), in which a product, CoA, is released between the binding of the two substrates. In the first and second step of the normal reaction the enzyme binds acetyl-CoA to a cysteine residue at the active site, rearranges and, after the expulsion of CoA, forms a stable, covalently bound acetyl-enzyme intermediate. Acetoacetyl-CoA, the second substrate, then binds to the acetyl-enzyme in the third step and rearranges in the fourth to give an enzyme-HMG-CoA intermediate which is hydrolysed in the fifth and final step to release the product and the catalyst (Figure 4).

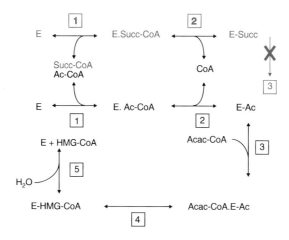

Figure 4. Reaction scheme for mitochondrial HMG-CoA synthase
Steps 1-5 represent the normal sequence of reaction steps catalysed by HMG-CoA synthase.
The scheme depicted in red represents the auto-succinylation and inhibition of the enzyme at the
active site, by succinyl-CoA. E, enzyme; Ac-CoA, acetyl-coenzyme A; E-Ac, acetyl-enzyme; Acac-
CoA, acetoacetyl-CoA; Succ-CoA, succinyl-CoA.

Lowe and Tubbs purified mitochondrial HMG-CoA synthase from ox
liver and characterized the enzyme. They found that HMG-CoA synthase is a
dimer of identical subunits; however, although each possesses an active site,
the enzyme only achieves half-of-sites reactivity. They showed that the purified synthase catalyses its own succinylation and inhibition by succinyl-CoA:
the enzyme binds succinyl-CoA at the active site, in place of acetyl-CoA, to
form a stable succinyl-intermediate, a dead-end complex, which cannot bind
the second substrate, acetoacetyl-CoA (Figure 4). They also found that purified HMG-CoA synthase desuccinylates with time and will bind acetyl-CoA,
the normal substrate, preferentially. Therefore, when presented succinyl-CoA
levels are low and the acetyl-CoA levels are high, the enzyme desuccinylates,
binds acetyl-CoA and is protected from resuccinylation. This observation led
them to propose that the reported glucagon-induced reduction of intramitochondrial succinyl-CoA levels might lead to desuccinylation and activation of
succinylated (inactivated) HMG-CoA synthase, and that such a mechanism
might be an explanation for the observed immediate and direct *in vivo* stimulation of hepatic ketogenesis by glucagon[17]. Their hypothesis was based solely
on experiments *in vitro* with purified enzyme and the observations of others
who had previously described the enzyme as 'rate-limiting for ketogenesis'[17].
More recent observations confirm that intramitochondrial succinyl-CoA levels
are of the range to effect inhibition of the enzyme[11] and that
desuccinylation/resuccinylation is a glucagon-induced, insulin-independent
control mechanism of mitochondrial HMG-CoA synthase in the rat[14], ox,
sheep and horse *in vivo* (P.A. Quant, unpublished work).

This control mechanism is unique. To date, it represents the only example
of covalent modification of an enzyme, at the active site, by a substrate analogue.

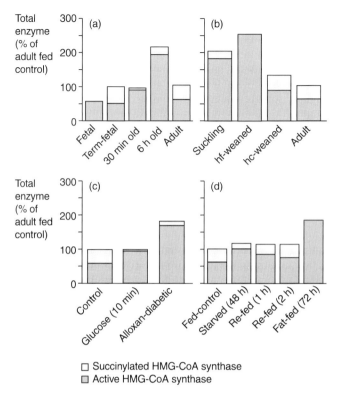

Figure 5. Effects of development, nutrition and hormones on succinylation state and absolute amounts of HMG-CoA synthase expressed in isolated rat liver mitochondria during certain transition periods
(a) fetal-neonatal transition; (b) suckling-weaning transition (c) onset of diabetes and (d) starved-to-fed transition.

In the rat, mitochondrial HMG-CoA synthase is controlled *in vivo* by two mechanisms: a rapid (minutes) fine-control mechanism, involving covalent modification of existing enzyme [succinylation (inhibition)/desuccinylation (activation)][14] and a longer-term (hours) control which changes the absolute amounts of enzyme[9] (Figure 5). Both these mechanisms control HMG-CoA synthase (in the rat) *in vivo* during the fetal-neonatal and suckling-weaning transition periods[9] (Figure 5).

Studies of the ontogeny of (rat) fetal mitochondrial HMG-CoA synthase have shown that total enzyme increases during late gestation (after the 18th day) to reach adult levels at term (22 days)[9] (Figure 5a). During the final 2 days before birth, HMG-CoA synthase activities do not increase further because the significant increase in the amounts of enzyme during this period is compensated for by succinylation of the same proportion of the enzyme. This provides a reservoir of inactivated enzyme which can be rapidly desuccinylated and reactivated immediately after birth, in response to the glucagon surge induced by the birth trauma and, therefore, a mechanism for increasing the capacity of the (HMG-CoA) ketogenic pathway. After birth, however, the absolute amounts of HMG-CoA synthase in the neonate are approximately twice those found in the adult rat and remain high during the suckling period[9] (Figure 5b) when, because glucagon levels are high and insulin levels low, most of the enzyme is active. Glucagon levels and HMG-CoA synthase activity fall with weaning on to the usual high-carbohydrate (hc) diet, but both remain elevated with weaning on to a high-fat (hf) diet[1,9] (Figure 5b); therefore, the proportion of enzyme which is succinylated (inactivated) is greater following high-carbohydrate rather than high-fat weaning.

Experimentally induced alloxan diabetes (Figure 5c), fat-feeding and starvation (Figure 5d) in adults also induce desuccinylation and increased amounts of the enzyme, while refeeding causes the opposite (Figure 5d). Thus the absolute amounts and the relative activity of the enzyme appear to vary with hormonal and nutritional status throughout development and with certain diseases in concert with ketogenic capacity[1,5,7,9,12-14,18] (Figure 5).

Therefore, as changes in ketone body production in mitochondria isolated from adult rats[11,12,14] and young rats[1,9,18] parallel changes in mitochondrial HMG-CoA synthase activity more closely than changes in CPT I activity, the question 'Does mitochondrial HMG-CoA synthase exert significant control over ketogenic flux?' arises naturally. Application of modern metabolic control analysis (Chapter 1) to the system is required to describe quantitatively how control over ketogenic flux is distributed[5,7,13] among these and other sites.

Application of metabolic control analysis to ketogenesis

Metabolic control analysis, devised by Kacser and Burns and Heinrich and Rapoport, and described fully in Chapter 1, enables quantitative study of control over flux, or over metabolites, by individual enzymes of a pathway. It provides answers to questions such as: 'How important is a selected individual enzyme in controlling pathway flux?' In this type of analysis, the importance of each individual step in a pathway in controlling the flow through that pathway in a steady state is given a numerical value, the flux control coefficient. This is defined as the fractional change in the pathway flux for an infinitesimal fractional change in the activity of a particular enzyme under study, with intermediates in the defined system allowed to relax to a new steady state. However, its application is restricted by the ability to manipulate individual enzyme activities, using specific inhibitors or genetic approaches, or to measure large numbers of elasticity coefficients[13]. As a full description of control structure can be built up from repeated applications of metabolic control analysis to different enzymes in the pathway, it is sometimes described as the 'bottom-up' approach[7,13,20].

Reasons for the lack of application of a full metabolic control analysis may include the complexity of the pathway (Figures 1, 2 and 3), [with acetyl-CoA and CoA involved at several points (Figure 1)], which makes it difficult to measure elasticity coefficients (measures of the responsiveness of enzymes, to changes in their effectors, which can be used to calculate flux control coefficients[7,13]) and the lack of suitable specific inhibitors of the intramitochondrial enzymes involved.

However, a newer approach to control analysis has been devised. In contrast, the 'top-down' approach to control analysis[20] circumvents many of the experimental limitations of the bottom-up control analysis as it does not rely on the use of specific inhibitors or genetic manipulation to determine control coefficients. It gives an immediate overview of the control structure of the whole pathway of interest by quantitatively describing the distribution of control among large blocks of reactions or branches of a pathway as 'group' or 'overall' flux control coefficients. It answers questions of the general nature: 'How is control over pathway flux distributed among the branches of the pathway?' or 'What control over pathway flux is exerted by groups of reactions, probably comprising many enzymes?'.

To use this approach the 'system' under investigation is conceptually divided into those reactions that produce and those that consume a suitable pathway 'intermediate'. Figure 6 shows how top-down control analysis can be applied to fatty acid oxidation and ketogenesis. If acetyl-CoA is the chosen intermediate and a long-chain fatty acid is used as the substrate, the group of enzymes producing acetyl-CoA would include: CPT I; the carnitine carrier; CPT II; the enzymes of β-oxidation; the enzymes of the electron transport chain and ATPase. If a medium- or short-chain fatty acid is used, then the mitochondrial-activating enzymes would be included instead of CPT I, the carnitine carrier and CPT II. The acetyl-CoA consumers would include the enzymes of the HMG-CoA cycle and of the Krebs' cycle, and the pyruvate carrier (Figures 1, 6).

By varying the concentration of acetyl-CoA and measuring the flux through the pathway, the group or overall elasticities of the acetyl-CoA producers and the acetyl-CoA consumers to acetyl-CoA can be calculated. (The group elasticities are complex functions of the elasticities of the individual component enzymes of the two blocks and measures of the sensitivities of the two groups of enzymes to changes in the concentration of the chosen intermediate, acetyl-CoA.) The level of acetyl-CoA can be manipulated by introducing a new pathway which is not part of the defined system under study. For this system, pyruvate can be added to change the acetyl-CoA concentration (Figure 6). Then, by using the summation and the connectivity theorems of control analysis (see Chapter 1), the distribution of control (between the two blocks of reactions either side of the intermediate) over this simple system can be solved. This is achieved by calculation from the two overall elasticities of the two group flux control coefficients. (These are sums of the individual flux control coefficients for each of the component enzymes in the two groups of reactions and numerical expressions of the control strength exerted by the two blocks of enzymes over the pathway flux.) For a full description of the general method and its application to this system in particular see [7,13,20].

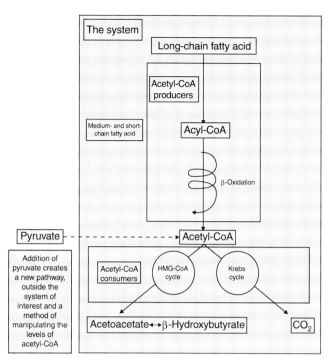

Figure 6. Application of top-down control analysis to fatty acid oxidation and ketogenesis
To apply top-down control analysis to fatty acid oxidation and ketogenesis the system is conceptually divided into two blocks: those reactions producing and those consuming the intermediate, acetyl-CoA. A new pathway is introduced, outside the system, to manipulate the levels of acetyl-CoA.

By application of this newer approach to control analysis in mitochondria isolated from rat liver, it has been established that the group of enzymes of the HMG-CoA pathway (Figures 1, 3 and 6), which include mitochondrial HMG-CoA synthase, can contribute significantly to (intramitochondrial) control of ketogenesis. These results question the frequent assumption that CPT I is necessarily the only major control site within the liver. Fuller analyses in isolated hepatocytes, whole organs and whole animals are now underway.

Implications and conclusions?

The results of these preliminary control analyses only apply to the contrived systems studied; however, the implications for human and veterinary medicine are far-reaching.

- *Preterm human infants do not appear to be able to produce and metabolize ketone bodies like full-term infants, and therefore lack the normal safety mechanism available when blood glucose fuel levels drop dangerously low[19]. Recently, it has been observed that polyclonal rabbit anti-ox-liver mitochondrial HMG-CoA synthase antibodies cross-react with the human enzyme. Preliminary observations have shown that the enzyme is demonstrable in fetal and infant liver samples, and a study of the ontogeny of the enzyme in the human fetus, infant and adult, and of its succinylation state under various dietary and endocrine conditions is being made[21]. This is providing important information on the ontogeny of ketogenic regulation under normal and pathological circumstances in the perinatal period. The lack of stimulation of ketogenesis, after the birth of premature infants, can be explained in part by reduced expression of HMG-CoA synthase protein and may also be due to lack of reactivation of succinylated (inhibited) enzyme. It is possible that informed choice (nature and percentage of fat) of diet could induce expression and activation of HMG-CoA synthase and relieve this problem[18,21].*

- *Lack of ketogenic capacity in suckling pigs can be explained by lack of mitochondrial HMG-CoA synthase[22], whereas control of ketogenesis appears to be different in horses and dogs (P.A. Quant, unpublished work).*

- *Succinylation of HMG-CoA synthase occurs in ruminants and may contribute to control of ketogenesis from medium- and short-chain fatty acids, which contribute significantly to ruminant metabolism (P.A. Quant, unpublished work).*

Conclusion

It is clear that regulation of ketogenesis is complex and cannot be explained by a single 'rate-limiting' step. Results of preliminary control analyses suggest that the role of mitochondrial HMG-CoA synthase is significant.

Suggestions for further reading

Fatty acid oxidation and ketogenesis

(Although the following two books are now somewhat out of date they are probably still the best sources of basic background information on metabolism.)

- Newsholme, E.A. and Leech, A.R. (1983) *Biochemistry for the Medical Sciences*. John Wiley and Sons, Chichester, New York
- Martin, B.R. (1987) *Metabolic Regulation: a Molecular Approach*. Blackwell Scientific Publications, Oxford, U.K.

Metabolic control analysis

(This book is not for the faint-hearted!)

- Cornish-Bowden, A. and Cárdenas, M.L. (eds.) *Control of Metabolic Processes* (1990) NATO ASI Series **190**, Plenum Press, New York, U.S.A.

References

1.   Girard, J., Ferré, P. Pégorier, J.-P. & Duée, P.-H. (1992) Adaptations of glucose and fatty acid metabolism during perinatal period and suckling-weaning transition. *Physiol. Rev.* **72**, 507-562

2.   Grantham, B.D. & Zammit, V.A. (1986) Restoration of the properties of carnitine palmitoyltrans-
      ferase I in liver mitochondria during re-feeding of starved rats. *Biochem. J.* **239**, 485-488
3.   Nehlig A. & Pereira de Vasconcelos, A. (1992) Glucose and ketone body utilization by the brain
      of neonatal rats. *Prog. Neurobiol.* **40**, 163-221
4.   Zammit, V.A. (1981) Intrahepatic regulation of ketogenesis. *Trends Biochem. Sci.* **6**, 46-49
5.   Quant, P.A., Robin, D., Robin, P., Ferré, P., Girard, J. & Brand, M.D. (1989) Control of aceto-
      acetate production from exogenous palmitoyl-CoA in isolated rat liver mitochondria. *Biochem.
      Soc. Trans.* **17**, 1089-1090
6.   Kunz, W.S. (1991) Application of the steady-state flux control to mitochondrial ß-oxidation.
      Biomed. Biochim. Acta **50**, 1143-1157
7.   Quant, P.A., Robin, D., Robin, P., Girard, J. & Brand, M.D. (1993) A top-down control analysis in
      isolated rat liver mitochondria: Can the 3-hydroxy-3-methylglutaryl-CoA pathway be rate-con-
      trolling for ketogenesis? *Biochim. Biophys. Acta* **1156**, 135-143
8.   Zammit, V. (1984) Mechanisms of regulation of the partition of fatty acids between oxidation and
      esterification in the liver. *Prog. Lipid Res.* **23**, 39-67
9.   Quant, P.A., Robin, D., Robin, P., Ferré, P., Brand, M.D. & Girard, J. (1991) Control of hepatic
      mitochondrial 3-hydroxy-3-methylglutaryl-CoA synthase during the foetal-neonatal transition,
      suckling and weaning in the rat. *Eur. J. Biochem.* **195**, 449-454
10.   Grantham, B.D. & Zammit, V.A. (1988) Role of carnitine palmitoyltransferase I in the regulation of
      hepatic ketogenesis during the onset and reversal of chronic diabetes. *Biochem. J.* **249**, 409-414
11.   Quant, P.A., Tubbs, P.K. & Brand, M.D. (1989) Treatment of rats with glucagon or mannoheptu-
      lose increases mitochondrial 3-hydroxy-3-methylglutaryl-CoA synthase activity and decreases
      succinyl-CoA content in liver. *Biochem. J.* **262**, 159-164
12.   Quant, P.A. (1990) Activity and expression of hepatic mitochondrial 3-hydroxy-3-methylglutaryl-
      CoA synthase during the starved-to-fed transition. *Biochem. Soc. Trans.* **18**, 994-995
13.   Quant, P.A. (1993) Experimental application of top-down control analysis to metabolic systems.
      Trends Biochem. Sci. **18**, 26-30
14.   Quant, P.A., Tubbs, P.K. & Brand, M.D. (1990) Glucagon activates mitochondrial 3-hydroxy-3-
      methylglutaryl-CoA synthase activity *in vivo* by decreasing the extent of succinylation of the
      enzyme. *Eur. J. Biochem.* **187**, 169-174
15.   Foster, D.W. (1984) Banting Lecture 1984: From glycogen to ketones--and back. *Diabetes* **33**,
      1188-1199
16.   Guzmàn, M. & Geelen, M.J.H. (1988) Short-term regulation of carnitine palmitoyltransferase activ-
      ity in isolated rat hepatocytes. *Biochem. Biophys. Res. Commun.* **151**, 781-787
17.   Lowe, D. & Tubbs, P.K. (1985) Succinylation and inactivation of 3-hydroxy-3-methylglutaryl-CoA
      synthase by succinyl-CoA and its possible relevance to the control of ketogenesis. *Biochem. J.*
      232, 37-42
18.   Thumelin, S., Forestier, M., Girard, J. & Pégorier, J.P. (1993) Developmental changes in mitochon-
      drial 3-hydroxy-3-methyl-glutaryl-CoA synthase gene expression in rat liver, intestine and kidney.
      Biochem. J. **292**, 493-496.
19.   Hawdon, J.M., Ward Platt, M.P. & Aynsley-Green, A. (1992) Patterns of metabolic adaptation for
      preterm and term infants in the first neonatal week. *Arch. Dis. Child.* **67**, 357-365
20.   Brown, G.C., Hafner, R. & Brand, M.D. (1990) A 'top-down' approach to the determination of
      control coefficients in metabolic control theory. *Eur. J. Biochem.* **188**, 321-325
21.   Quant, P.A., Henson, J.N., Williams, A., Jeffrey, I., Taylor, P. & Carter, N.D. (1993) Does reduced
      expression of 3-hydroxy-3-methyl-glutaryl-CoA synthase in preterm infants explain lack of stimu-
      lation of ketogenesis following birth? *J. Physiol.* **467**, 367P
22.   Dueé, P.-H., Pégorier, J.-P., Herbin, C., Quant, P.A., Kohl, C. and Girard, J. (1994) *Biochem. J.* **298**,
      207-212.

3

Motor neurone disease

Ajith Goonetilleke*, Jacqueline de Belleroche† and Roberto J. Guiloff*

Departments of Neurology* and Biochemistry†, Charing Cross Hospital and Charing Cross and Westminster Medical School, London, U.K.

Introduction

Motor neurone (MN) disease or amyotrophic lateral sclerosis (ALS), is a severe progressive neurological disorder with an incidence of 1-1.5/100 000 /year and a prevalence of 4-8/100 000. Peak age of onset is in the 6th decade, with a male/female ratio of approximately 2:1.

Charcot first outlined the clinical and pathological features of classical or sporadic ALS in 1869, and placed emphasis on the degeneration of upper and lower motor neurones (UMNs and LMNs). A number of pathological and biochemical abnormalities have been identified since, but the cause(s) and pathogenesis remain obscure. No single hypothesis of the aetiopathogenesis of ALS has been accepted and no treatment has been shown to convincingly modify the course of the disease.

This chapter will describe the main clinical, neuropathological and biochemical features of classical ALS. Atypical (Western Pacific, juvenile) and familial forms are discussed with proposed aetiological factors and hypotheses on pathogenesis. Treatment options will also be considered.

Clinical features

UMNs connect the cerebral cortex to the brain stem and spinal cord. LMNs extend from the brain stem to the muscles of eyes, jaw, face, pharynx/larynx and tongue, and from the spinal cord to the diaphragm and muscles of the trunk, limbs and sphincters. ALS results in muscle weakness, and is characterized by clinical signs of involvement of UMNs (spasticity, increased tendon reflexes, extensor plantar responses) and LMNs (muscle wasting, fasciculations) (Figure 1). The limbs, trunk, respiration, speech and swallowing can be involved. In contrast, disturbances in intellectual function, ocular movements, bladder/bowel function and sensation are rare. The condition is invariably fatal, with a 75% mortality within 5 years of clinical onset. Sometimes patients may survive for 10 or more years. The cause of death is usually respiratory failure due to chest infection.

Classical or sporadic ALS

The most frequent clinical form is characterized by the onset of UMN and LMN involvement in the limbs. Bulbar features are common; when UMN and LMN involvement of the brain stem territory predominates from the onset the term 'bulbar palsy' is used. Clinical presentations with only LMN (progressive muscular atrophy, PMA) or UMN signs are rarer. In both instances, a definitive diagnosis of ALS can only be sustained by the subsequent clinical or pathological demonstration of UMN (in PMA) and LMN involvement.

A diagnosis of ALS requires the exclusion of other diseases that produce UMN and/or LMN signs, for example the more benign forms of adult spinal muscular atrophy, which show slow progression and do not have UMN signs. The 'El Escorial' criteria, now accepted by the World Federation of Neurology, classify a clinical diagnosis of ALS as 'definite' if progressive UMN and LMN signs are present in three or four territories (e.g. bulbar, trunk, upper and lower limbs). A number of diseases can mimic ALS by producing UMN and/or LMN signs. They include brain tumours, strokes and herniated cervical and lumbar discs pressing either on the spinal cord or the nerve roots. Methods that can image the nervous system are useful; they include magnetic resonance imaging (m.r.i.) and computerized axial tomography (c.a.t. scan) combined with myelography (injection of a radio-opaque substance in the spinal canal). Electromyography is used to confirm denervation (i.e. LMN involvement) in clinically affected muscles, and to establish it in others which may appear normal. Nerve conduction studies are useful to exclude peripheral nerve disorders capable of producing muscle denervation and wasting. The cerebrospinal fluid (CSF), obtained by a lumbar puncture, should be normal in sporadic ALS, or may have a slightly raised protein level. Other investigations may be used to exclude conditions such as diabetes (blood sugar), thyrotoxicosis (thyroid hormone levels), paraproteinaemia (protein electrophoresis), syphilis (serological tests) or lymphoma (tissue biopsy).

Neuropathology

ALS is characterized by loss of MNs with foci of astrocytic gliosis and degeneration of the pyramidal tracts (Figure 2). In absolute numbers, the loss of MNs is most prominent in the spinal cord and brain stem (LMNs), but is also

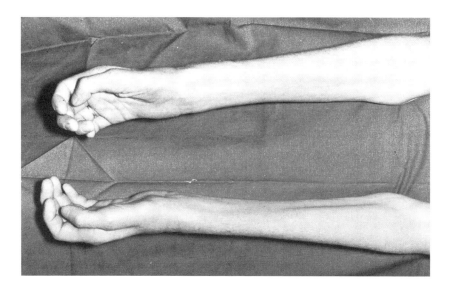

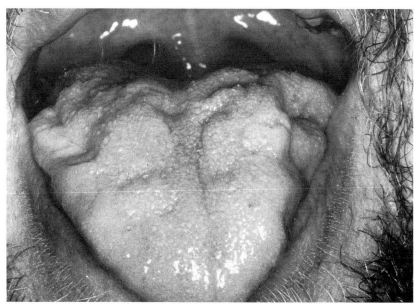

Figure 1. LMN signs in a patient with ALS
Top: wasting of arms and hands (cervical cord territory).
Bottom: wasting of the tongue (bulbar territory).

found in the motor cortex (UMNs). Brainstem motor nuclei V, VII, nucleus ambiguus and XII, which supply muscles of jaw, face, pharynx/larynx and tongue, respectively, are most susceptible. There is preservation of motor nuclei III, IV and VI, which supply muscles mediating ocular movements, except in the very late stages of the disease (e.g. patients kept alive with respiratory support machines for several years). LMNs of Onuf's nucleus (situated in

(a)   (b)

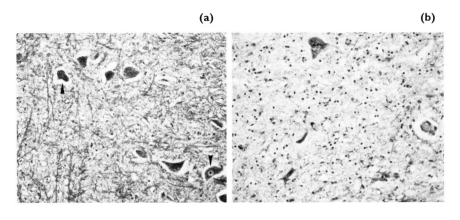

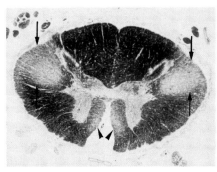

Figure 2. Pathology of ALS
Photomicrograph of the hypoglossal nucleus, where MNs innervating the tongue originate, in (a) normal and (b) ALS. Note the striking reduction in the number of MNs (arrowheads). (c) Transverse cut (stained for myelin) through the spinal cord in a case of classical ALS. Note pale areas in the lateral (arrows) and anterior columns (arrowheads) due to degeneration of the pyramidal tracts.

the sacral part of the spinal cord), which supply the bladder and rectum, are curiously spared. The degeneration of the pyramidal tracts is most marked in the lateral and anterior columns of the spinal cord, but can be found also in the brain stem and in the hemispheres of the brain. The topography of pathological changes in ALS accounts for the clinical features.

Using light microscopy, dendritic atrophy in MNs is an early finding in ALS. Surviving MNs show increased concentrations of lipofuscin granules, which are pigments usually found in ageing neurones. Changes in nuclear and cytoplasmic components of affected neurones are also present. Staining for nuclear chromatin (DNA) shows a diffuse pattern in normally active cells. There is chromatin clumping in ALS, and in severely affected neurones the nucleus is usually shrunken. Nissl granules (RNA), which are in the cytoplasm, are often lost or aggregate into masses at the cell periphery in ALS.

MN inclusions can be identified in ALS by their appearance and staining characteristics. Some are characteristic and found in the cytoplasm, like Bunina (small, round or oval, eosinophilic), hyaline and basophilic bodies. Large proximal axonal non-specific inclusions (spheroids) can also be seen. Other non-specific inclusions include Hirano bodies and Lewy body-like inclusions, the latter are similar to those seen in Parkinson's disease. A number of the above inclusions are ubiquitin immunoreactive and are characteristic markers of the disease. Ubiquitin is a low molecular mass protein implicated in the degradation of some cellular proteins; studies in neurodegenerative disorders have demonstrated ubiquitin immunoreactivity in neurofibrillary tangles and senile plaque neurites (Alzheimer's), Lewy bodies (Parkinson's) and Pick bodies (Pick's disease). Ubiquitin immunoreactivity in ALS occurs as fine (10-15 nm) or coarse (15-20 nm) skeins of filamentous material, or as dense bodies, in the cytoplasm of MNs. Fine skeins are more common and also seen in apparently normal MNs[1]. Although they are not labelled by antibodies to components of cytoskeletal proteins (as in other neurodegenerative disorders), it is thought that ubiquitin accumulation in ALS results from its conjugation to abnormal proteins that are resistant to degradation by this pathway.

Biochemical pathology

Neurotransmitter changes

A number of abnormalities have been found in ALS. In the spinal cord and brain stem, LMN loss is associated with a decrease in choline acetyltransferase (ChAT) activity and ChAT mRNA (see below). Few other consistent changes are seen in the spinal cord, but decreases in substance P and calcitonin-gene-related peptide immunoreactivity, thyrotrophin-releasing hormone (TRH) and serotonin (5-HT) are found. Substantial decreases in muscarinic acetylcholine receptors were found by Whitehouse[2] but not Hayashi[3], who found a decrease in strychnine-sensitive glycine receptors in anterior grey matter, but not in muscarinic, dopaminergic, γ-amino butyric acid (GABA) and β-adrenoceptor binding. In the ventral horns there is a decrease in TRH receptors and an unexplained increase in $5-HT_{1A}$ receptors.

UMNs in ALS motor cortex have been more difficult to study. The transmitter in this pathway is glutamate, which is responsible for rapid excitatory responses through ligand-gated ion channel receptors. There have been claims that glutamate levels in the plasma and CSF of ALS patients are elevated. However, these results have been difficult to reproduce. Glutamate CSF levels are particularly low and appropriate storage of CSF samples is necessary to prevent glutamine breakdown to glutamate. This has, however, led to speculation that excitotoxic levels of glutamate might be present in ALS (see below)[4]. Support has come from the report of reduced glutamate uptake in ALS spinal cord and cerebral cortex[5]. The glutamate transporter plays a vital role in the termination of the action of glutamate at the synapse. However, the reduced binding of glutamate by the synaptosomal membrane found may be due to factors other than a reduced transporter. We have found elevated CSF glycine levels in ALS which, if the glutamate transporter is depleted, would potentiate glutamate-mediated neurotoxicity at the N-methyl-D-aspartate (NMDA) glutamate receptor subtype. Glycine functions as a co-agonist with glutamate at the NMDA receptor and potentiates both the physiological and neurotoxic actions of glutamate.

Gene expression in ALS

Several studies indicate that impaired transcription occurs in MNs in ALS. Nuclear and nucleolar volumes and RNA content, determined both in tissue extracts and by microdensitometry, are reduced. The RNA content of ALS cells appears to be decreased in both histologically normal and abnormal cells[6]. This has provided a basis for studying the expression of individual genes that may give insight into the aetiology of ALS. However, no change in the expression of neurofilament light-chain mRNA, an abundant neuronal intermediate for the protein, was detected in MNs studied by *in situ* hybridization, and no changes in the neuronal mRNA encoding the amyloid precursor protein (APP) have been detected in tissue extracts by Northern analysis[7]. However, using an oligonucleotide probe specific for ChAT mRNA, we have found a differential loss of ChAT mRNA in anterior horn cells in ALS compared with control spinal cords[8]. This change occurs without a parallel loss in other mRNAs, such as G-protein (guanine-nucleotide-binding protein) α-subunit mRNA in adjacent sections (Figure 3). Marked loss of ChAT mRNA was also

(a)   (b)

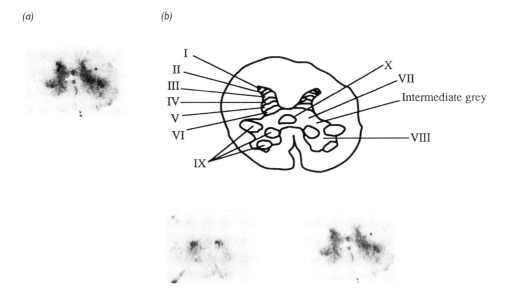

Figure 3. Gene expression in ALS
Distribution of ChAT mRNA in human spinal cord determined by *in situ* hybridization. The distribution of ChAT mRNA in control spinal cord (a) is compared with the laminae defined by cytoarchitecture (b). The lower Figures show the loss of ChAT mRNA in ventral grey matter of MN disease spinal cord (left) compared with the control tissue (right). This Figure is reproduced from Virgo *et al.* 8, with permission from Elsevier Science publishers NV.

seen in regions containing sensory neurones and in the central canal cells of a number of cases. This is consistent with the loss of muscarinic receptors and of ChAT activity[9] seen in dorsal horn in ALS, further indicating that the changes in ALS may be extensive and not solely restricted to the MNs.

Studies on the expression of GAP-43 mRNA, the growth-associated protein, show that this mRNA is increased in anterior horn cells in ALS[10]. However, the post-mortem delay in freezing for controls was twice as long as that for ALS patients. Such an increase, if confirmed, may represent a mechanism for protecting neurones from degeneration or contribute to compensatory axonal sprouting from spared MNs to denervated muscle fibres.

Aetiology

Sporadic ALS constitutes about 95% of cases of ALS worldwide. We review here the main proposed aetiological factors and hypotheses on the pathogenesis of ALS. Unfortunately, there is no satisfactory animal model for classical ALS. The existence of a familial form makes it possible to approach the mechanisms of MN degeneration by identifying the gene(s) responsible for this form, in the hope that the chain of events induced by such gene(s) is similar to that operating in classical ALS. The high incidence of atypical forms of ALS in certain foci in the Western Pacific has prompted much work on possible environmental causes. Therefore, these forms are considered in this section.

Aetiological factors

Familial amyotrophic lateral sclerosis (FALS)

Of ALS cases, 5-10% are familial (Figure 4), most showing autosomal dominant inheritance with a mean age of onset of 50 years and duration of illness of 3 years. Variable expression of the disease is seen within single families, with a variation of 40 years in the age of onset and from 6 months to 5 years in duration of illness. Signs at onset can also vary as can the degree of involvement of UMN. In some studies, familial forms of ALS show a later age of onset (76 years) and many obligate carriers in these families die before manifesting the disease. Isolated cases within a family may show features of dementia or Parkinsonism but this is not common. Autosomal recessive and juvenile onset forms of the disease have also been reported. A large number of families in North America and Europe have been used for linkage analysis.

Linkage analysis in FALS

Linkage between a specific genomic marker allele and the disease is calculated by maximum likelihood analysis. The likelihood that the marker and disease locus are close together or linked depends on the frequency of recombination events which separate the loci (and hence the specific allele) associated with the disease. The closer the loci, the more frequently they co-segregate. The pro-

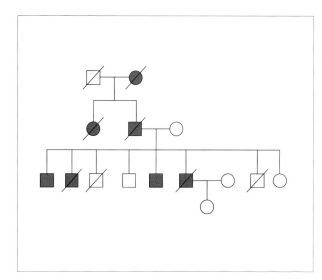

Figure 4. Familial ALS (FALS)
Pedigree of a U.K. family with autosomal dominant inheritance. Squares, males; circles, females; filled, affected; unfilled, unaffected; diagonal line, dead.

portion of recombinants is called the recombination fraction, θ (theta). For unlinked loci, θ = 0.5 (50% recombination). The likelihood method of analysis estimates θ, and tests whether this is significantly smaller than 50%. The likelihood (odds) that the two loci are linked is calculated as an odds ratio, this being the ratio of the odds in favour of the loci being linked to the odds in favour of free recombination (50% recombination). For convenience this is expressed as the $\log_{10}$ of the odds ratio, or lod score. Significant linkage is established when lod scores of >+3 (i.e. more than 1000 to 1 in favour of linkage) are obtained; lod scores <–2 exclude linkage. It is also useful to calculate multipoint lod scores using multiple markers in close vicinity to one another on the chromosome.

DNA from familial cases has been extensively screened on Southern blots for restriction fragment length polymorphisms (RFLPs), and several areas in the genome have been excluded. There is evidence of linkage between the disease locus and chromosome 21 loci in a subset of families[11]. This result was obtained initially from multipoint analysis using four markers, D21S52, D21S1/S11, APP and D21S58 which yielded a significant lod score of 5.03 (i.e. odds of 100 000 to 1 in favour of linkage) at 10 centimorgans (cM) telomeric to D21S58.

Recently, emphasis has focused on the use of polymorphisms in highly variable regions which possess multiple alleles at each locus, in contrast to the use of RFLP probes which detect diallelic polymorphisms. A large number of polymorphic loci containing dinucleotide (CA) repeats have now been identified (microsatellite polymorphism), which has enabled a reasonably comprehensive linkage map of the human genome to be constructed. There are 605 loci with a heterozygosity greater than 0.7, and 553 that can be ordered with an odds ratio greater than 1000:1[12]. Ideally, markers should be evenly spaced 2-5 cM apart, but at present the average distance is 5 cM.

The publication of linkage to chromosome 21 markers prompted us to screen families in the U.K. for chromosome 21 microsatellite polymorphisms. We used eight such polymorphisms flanking the putative FALS locus, in order from the centromere: D21S120, D21S214, D21S210, IFNAR, D21S223, D21S224, D21S225 and D21S156. Multipoint analysis of these five loci on eight families gave a significant exclusion of the FALS locus from a genetic distance of approximately 13.5 cM around D21S214 and D21S210, and 58 cM covering a region extending from IFNAR to D21S156, with no significant linkage or exclusion at D21S210[13]. Clearly at least two loci may be responsible for FALS. This is comparable to hereditary sensory and motor neuropathy type 1 (Charcot-Marie-Tooth disease), where abnormalities of three loci on chromosomes 1, 17 and X produce a similar clinical phenotype.

Recently, significant two-point lod scores (4.25 at a recombination fraction of 0.1) were reported for D21S223 in FALS families[14]. As D21S223 is located close to the Cu/Zn superoxide dismutase gene (SOD1), both being present in the same chromosome 21 cosmid, the presence of mutations in this gene have been investigated. Initially, single-strand conformational polymorphism (SSCP) analysis has detected mobility shifts of single-strand DNA caused by sequence variations. Data from the analysis of two of the five exons (exons 2 and 4) of SOD1 revealed polymorphisms in 6/15 families where the FALS locus was linked to chromosome 21 markers and in 12/135 other families, with no abnormalities in 140 controls[14]. Further confirmation was carried out by sequence analysis of the SOD1 exons in 13 of the 18 families with anomalous SSCP bands and this showed the presence of single base changes in the SOD1 gene. Further studies have identified mutations in exons 1 and 5 and demonstrated that mutations are associated with a significant reduction in SOD1 enzyme activity. More work is needed to investigate what potential mechanisms may exist for the development of FALS. If an abnormality in SOD1 proves to be the underlying cause of some chromosome-21-linked families, treatment through free-radical regulation may prove useful, but non-chromosome 21 loci still need to be identified.

Western Pacific ALS (WP-ALS)

Incidence rates of ALS 50-100 times higher than elsewhere were found in the 1950s in the Chamorro people of the southern Marianas Islands (Guam is the largest), and later in the Auyu and Jakai people of West New Guinea and in the inhabitants of the Kii peninsula of Japan. Clinically ALS is usually combined with features of Parkinson's disease and/or dementia (ALS-PD complex). The pathology differs from classical ALS; extensive neurofibrillary tangles (NFTs) are seen in hippocampal, cortical and brainstem areas. Multinucleated cells in the cerebellum (Purkinje cells) and other brain areas have been described, as has the presence of Purkinje cells in other layers of the cerebellum than usually found. These findings suggest abnormal neuronal mitoses and migration during early development. Paired helical filaments

(PHFs) characterize the NFTs, as in Alzheimer's disease, but without the senile plaques seen in the latter. The significance of NFTs in WP-ALS is unclear, as they are also present in Guam cases of pure Parkinson's disease and, as an age-related finding, in normal Guamanians.

An environmental cause has been suggested in WP-ALS because (i) the three foci have geographically and genetically distinct populations, (ii) there was a 5-10-fold increase in incidence of ALS in long-term Filipino migrants to Guam and in Guamanian Chamorro migrants to the US mainland, and (iii) the incidence of ALS-PD in all three foci has steadily come down since 1960. Migrant studies suggest a latency of 18-20 years from exposure in Guam and onset of ALS.

Cycads

The plant *Cycas circinalis* is abundant in the Western Pacific. Cattle and sheep grazing on cycads can develop a progressive and irreversible paralysis of hindlimbs, with preservation of bladder, anus and tail function. Washed and soaked cycad seeds were a traditional source of flour, whole dried cycad seeds were used as medicines, and fresh cycad kernels used as a poultice for open wounds. The decline in WP-ALS coincides with Westernization of diet and medical practice. One of three monkeys fed with washed and soaked cycad seed for 9 months developed clinical and pathological features reminiscent of ALS.

Cycads have two potential neurotoxins (Figure 5): (i) cycasin (glycone of methylazoxy methanol) is a carcinogen, with hepato- and neurotoxicity. Newborn rodents exposed to high levels developed abnormally located Purkinje cells in the cerebellum, reminiscent of the changes described in WP-ALS (see above). (ii) β-*N*-methylamino-L-alanine (BMAA), fed to monkeys in large quantities for several weeks, produces a non-progressive motor disorder clinically and histologically dissimilar to WP-ALS[15]. There is too little BMAA in washed cycads to postulate a similar toxic effect in humans.

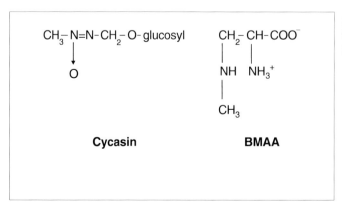

Figure 5. Neurotoxins from cycads
(a) Cycasin; (b) BMAA.

Soil abnormalities

A high incidence of ALS–PD was detected in 1965 among local manganese miners. Geochemical analyses in all three foci showed high levels of aluminium and manganese, with low levels of calcium and magnesium in soil and natural water supplies[16]. It was suggested that the decline in incidence of ALS–PD related to the introduction of public water systems. Central nervous system (CNS) tissues of WP-ALS in Guam and Kii show high levels of aluminium and calcium but not manganese, with selective deposition of aluminium in NFTs. It was hypothesized that chronic nutritional deficiencies of calcium and manganese resulted in secondary hyperparathyroidism, leading to increased absorption of aluminium from the gut and increased aluminium deposition in tissues, including cerebral grey matter. Primates fed on diets simulating the trace element disturbances on Guam have shown pathological changes in MNs, including NFTs. However, all animals remained clinically well for up to 4 years on these diets and the neurofilaments did not contain PHFs. Further, human hyperparathyroidism is not associated with ALS and elevated levels of aluminium in dialysis patients are not associated with ALS, but with dementia.

Trace elements

Consistent abnormalities of copper, zinc, lead or mercury levels have not been shown in studies on serum, cerebrospinal fluid and CNS tissues of ALS patients. This may relate to differences in preparation (e.g. unfixed/fixed cords) and analysis (e.g. neutron activation analysis, X-ray microanalysis, X-ray spectroscopy) of specimens in various laboratories. Also, disturbances in trace elements in ALS may be secondary to neuronal dysfunction or to alterations in the blood-brain barrier, rather than a primary abnormality.

Calcium, aluminium and manganese have been discussed for WP-ALS. High intracellular Ca^{2+} levels with generation of free radicals may be the common mechanism for cellular death. MNs might be more susceptible than other cells because of lower levels of intracellular Ca^{2+}-binding proteins such as parvalbumin and calbindin D28-K. Chronic Al^{3+} exposure has resulted, *in vitro* and *in vivo*, in intracellular accumulation of neurofilaments, possibly on the basis of an electrical charge effect of intraneuronal Al^{3+}. Primates fed high aluminium/manganese and low-calcium diets failed to show accumulation of manganese in the spinal cord. The results in ALS spinal cords have been conflicting.

Autoimmunity

Autoimmune attack of MNs is another possible mechanism of disease production. Reports of a higher than normal incidence of autoimmune disorders and paraproteinaemias in ALS have been conflicting.

Animal models of immune-mediated MN destruction

Two have been described[17]. Experimental autoimmune motor neurone disease

(EAMND) is an LMN syndrome induced in guinea-pigs by 5 monthly injections of purified bovine spinal cord MNs. Degenerating MNs have demonstrable IgG deposition in their cytoplasm, as well as at the neuromuscular junction. Experimental autoimmune grey matter disease (EAGMD) is a more acute disorder, affecting UMN and LMN, and is induced by two inoculations of spinal cord ventral horn homogenates. IgG deposits similar to those in EAMND occur early in the disease, followed later by deposition on the external membrane of pyramidal neurones. ALS spinal cords and motor cortex stained more intensely for IgG than controls. Although mice injected with sera from moderate/severe EAGMD developed weakness, this has not been reproduced with sera from ALS patients.

Ganglioside disorders

Gangliosides are lipid glycoconjugates found in high concentration in neural membranes (Figure 6).

Deficiency of hexosaminidase A (lysosomal enzyme that catalyses the degradation of GM2 gangliosides) can result in MN involvement, but not classic ALS. In one series, 78% of ALS patients had demonstrable antibodies to GM1 or GD1a. Using titres of greater than 1:1000 as positive, to ensure greater specificity, 19% of ALS were positive for anti-GM1, compared with 10% of other neurological diseases controls[18]. Most cases with high antibody titres have a peripheral neuropathy, not ALS. In conclusion, gangliosides are unlikely to be related to the aetiology of ALS.

Neurotrophic factors

These factors regulate the survival and development of certain neuronal populations. The first described was the nerve growth factor (NGF) a 13 kDa basic polypeptide made by target tissues of sensory and sympathetic neurones, as well as in the brain. NGF may be important in early life, when these neurones compete for a limited supply of NGF for survival. NGF may also have a role in adult life, as Schwann cells distal to a nerve crush express increased amounts of NGF and NGF-receptor, perhaps related to axonal regeneration. NGF is transported retrogradely from nerve terminals to the cell body. Brain-derived

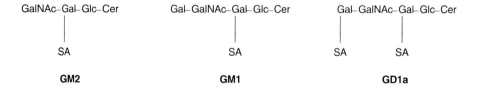

Figure 6. Gangliosides
GalNAc, N-acetylgalactosamine; Gal, galacatose; Glc, glucose; Cer, ceramide; SA, Sialic acid.

neurotrophic factor (BDNF) is mainly synthesized in the CNS, with neurotrophic effects on retinal ganglion cells and peripheral sensory neurones.
BDNF is also made by target cells, neurones becoming BDNF-dependent
once contact with target tissues has been made.

Survival of cultured MNs is promoted by many factors. Ciliary neurotrophic factor (CNTF), leukaemia-inhibiting factor and basic fibroblast
growth factor are the most potent; others such as insulin-like growth factor 1
(IGF1), transforming growth factor B and other fibroblast growth factors have
less activity. CNTF is a 22.8 kDa cytosolic molecule that also promotes MN
survival *in vivo*. In rats it is expressed exclusively postnatally (from the 4th
day), in Schwann cells and some sub-populations of glial cells. This, in addition to CNTF preventing MN degeneration after experimental axotomy in
early postnatal life, suggests a role more as a lesion factor than as a target-
derived neurotrophin. In mouse mutant progressive motor neuronopathy, an
autosomal recessive condition with MN degeneration, treatment with CNTF
may prolong survival, improve motor function and reduce loss of MNs[19]. To
date, no abnormalities of CNTF levels have been demonstrated in ALS
patients.

Androgens play a role in the regulation of the number and size of MNs
innervating perineal muscles during development, but androgen receptors are
also present in motor cranial nerve nuclei and in the spinal cord of the rat. It
has been suggested that neuronal loss in ALS may be secondary to loss of
androgen receptors[20]. X-linked spinal and bulbar muscular atrophy, an adult
onset form of LMN disease with gynecomastia and reduced fertility, suggesting androgen insensitivity, is caused by a mutation in the androgen receptor
gene[21]. Androgens may be important for adult LMN survival, but their role in
ALS is yet to be determined.

Viruses

Although there are some case reports of MN involvement with viral infections,
studies on populations of ALS patients have failed to implicate any single
virus. Poliomyelitis was suggested as a cause because ALS patients had a past
history of polio infection more often than expected[22]. However, many of those
patients did not have UMN signs, others have failed to confirm a higher incidence of prior polio infection and the poliovirus has not been isolated from
ALS tissues. Mumps was implicated when 20% of ALS patients were found to
have a past history of mumps infection in adulthood. Subsequent studies in
ALS have not shown elevated anti-mumps antibody titres nor isolated mumps
virus from tissues. The retrovirus human T-cell leukaemia type 1 (HTLV-1)
causes tropical spastic paraparesis which is clinically and pathologically different from ALS. A few cases of ALS have been reported in HIV1 seropositives,
but this may be a coincidence.

Trauma (mechanical, fracture, electrical or operative) has been associated
with the onset of ALS in several retrospective surveys. However, in any 12

month period a third of the general population may experience a trauma severe enough to be recalled at interview, and over 10 years about 98% may experience such events. The risk of recall bias in retrospective studies is apparent. A prospective study of back and head injuries has, to date, shown no association with ALS[23]. If trauma was related to ALS, the selectivity of damage to MNs will need to be explained.

Pathogenetic hypotheses

Excitotoxicity
High synaptic concentrations of excitatory amino acids can lead to death of postsynaptic neurones[24]. The abnormalities in glutamate metabolism reported in ALS led to the hypothesis that high levels of this excitatory transmitter in the synaptic cleft might result in damage to MNs following NMDA receptor activation[2]. This important hypothesis remains unproven, but the idea has been already tested in treatment regimes.

Work with glutamate analogues (Figure 7), such as β-N-oxalylamino-L-alanine (BOAA), present in the chickling pea *Lathyrus sativus* and causing UMN degeneration in lathyrism, suggests that non-NMDA receptors may also mediate MN damage[15]. BOAA binds to the glutamate receptor subtype also activated by the agonists α-amino-3-hydroxy-5-methyl-4-isoxazole (AMPA) and kainate.

Premature ageing of MN
This hypothesis[25] states that ALS, Parkinson's and Alzheimer's disease are due to environmental subclinical damage to specific neuronal populations, producing symptoms several decades later when superimposed on age-related neuronal death. Compensatory mechanisms could delay the onset of symptoms,

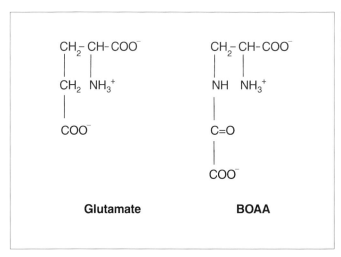

Figure 7.
Glutamate (left) and its analogue BOAA (right)

including neuronal reserve, re-innervation by axonal sprouts from surviving neurones, increased synthesis of neurotransmitters, increased number of post-synaptic receptors and improved performance of other postsynaptic structures. Individual susceptibility to environmental agents could be modified by endogenous (initial neuronal complement, detoxification capacity) or exogenous (diet, gut flora) factors.

Trans-synaptic degeneration

The presence of a combined loss of UMNs and LMNs in ALS has led some to consider that the degenerative process spreads across the synapses between them. This may account for the selective loss of MNs in this disease. The anterograde theory[26] suggests spread from UMNs to LMNs; degeneration of the LMN may be secondary to reduced neurotransmitter/growth factors from the presynaptic cell. Conversely, the retrograde theory[27] suggests spread from LMNs to UMNs; this may occur by a transfer of pathogens/toxins, or by a reduced supply of growth factors, from the postsynaptic cell. The anterograde theory needs to account for the paucity of reports of LMN involvement following UMN loss, as in strokes. The retrograde theory has the same difficulty explaining the lack of UMN involvement when widespread LMN destruction occurs, as in polio. Kiernan and Hudson[28] found no correlation between the degree of LMN loss in the spinal cord in ALS and the size of cortical MNs in the precentral cortex, suggesting independent degeneration.

Dendrites of MNs, as opposed to other neurones, contain prominent bundles of neurofilaments. Some have suggested that neurofilaments, probably involved in maintaining cytoskeletal structure, may be depleted in MN dendrites[29], impairing afferent inputs and resulting in gradual inactivity of MNs. This might lead to an interruption of supply of trophic substances, with resultant cell death. Retrograde axonal transport is decreased in ALS[30].

Abnormal MN DNA/RNA

Reduced levels of RNA in MNs of ALS and of the wobbler mouse (animal model) have been reported. This led to the hypothesis that the initial site of pathology in ALS was the nucleus, with the condensation of chromatin from an 'active' diffuse form to an 'inactive' clumped form. Inactive DNA would result in reduced transcription of RNA. As cell protein synthesis is dependent on RNA, affected cells would become metabolically compromised and die[31].

Accumulated DNA damage from genetic and/or environmental factors has been suggested as a cause of MN degeneration in ALS. If a primary defect in DNA repair enzymes existed, this would lead to disease. Mean survival and mean DNA synthesis in cultured ALS fibroblasts were found to be reduced by an alkylating agent compared with controls, but not by u.v. light, X-rays or mitomycin[32].

The changes described in RNA levels/DNA repair in ALS could be the result of disturbed function in already diseased cells, rather than constitute a step in the chain of events leading to MN death.

Failure of detoxification might result in neuronal death
A study in ALS subjects revealed impaired capacity for sulphur oxidation and conjugation, but normal carbon oxidation and glucuronide conjugation. Further work showed low plasma sulphate levels in ALS, Parkinson's and Alzheimer's disease[33]. Low inorganic sulphate levels, being the rate-limiting step in sulphoxidation, could lead to reduced xenobiotic detoxification in these subjects. Thus, ALS might result from a genetic predisposition (sulphoxidation may have recessive inheritance) and variable exposure to toxin(s). Poor metabolizers would require less toxin exposure to develop disease. The toxin(s) would have to act selectively on certain groups of neurones.

Treatment options

Autoimmunity has been implicated in the pathogenesis of ALS, but trials of plasmapheresis and immunosuppressive drugs have been disappointing. Transient improvements with immunosuppressants in some patients with motor peripheral neuropathies and very high anti-GM1 (ganglioside M1) titres have been reported.

The putative role for glutamate in MN loss in ALS was first tested in a trial of branched-chain amino acid treatment. The rationale for their use was their stimulatory effect on glutamate dehydrogenase, and hence on glutamate metabolism. However, the activity of this enzyme is normal in ALS and, despite initial optimistic claims, little support has since been obtained in other trials[34]. Trials have recently focused on reducing glutamate action with anti-epileptic drugs, but no conclusions have yet been reached. If NMDA or non-NMDA receptors mediated MN loss in ALS, then intervention with specific antagonists, or drugs acting pre- or postsynaptically, may be feasible; one such trial is currently in progress using riluzole (α-2-trifluoromethoxy-6-benzo-thiazole).

There is much interest in the use of neurotrophins in ALS. Evidence of trophic activity of TRH *in vitro* has led to trials in ALS. Long-term use of TRH is limited by its short half-life and poor oral bioavailability. Short-term studies of RX77368 (an analogue with a longer half-life than TRH) in ALS have shown beneficial effects on bulbar function[35]; long-term studies are in progress. Therapeutic trials with CNTF and IGF1 are currently underway in ALS patients. The rapid advances in neurobiological techniques suggest that a variety of other trophic factors for MNs may be discovered in the next few years; those with potential to enhance survival and protect the function of adult MNs would be the best candidates for clinical trials in ALS.

The work with FALS raises hopes of finding the responsible genes and of characterizing the chain of events that lead to UMN and LMN loss. If an abnormality in the SOD gene leads to MN death in ALS, then free-radical regulation might prove of therapeutic value.

Summary

- *Motor neurone disease, or amyotrophic lateral sclerosis, is a serious progressive neurological disorder, characterized by loss of UMN and LMN. Pathological features include characteristic intracytoplasmic MN inclusion bodies and appearances on ubiquitin staining. The aetiopathogenesis of the disease remains unknown and there is, to date, no effective treatment.*

- *Several abnormalities have been demonstrated in neurotransmitter, neuropeptide and gene expression studies. Abnormalities in glutamate metabolism have led to the excitotoxin hypothesis of MN destruction. Other theories include deficits in MN trophic factors, trans-synaptic degeneration, impaired ability to detoxify putative toxic agents and impaired DNA/RNA metabolism. The existence of familial forms, some of which show linkage to markers in chromosome 21, allows a genetic approach to the mechanisms of disease. Recent studies suggest that mutations in the Cu/Zn SOD gene may be important in some of the familial forms. The atypical forms seen in the Western Pacific have stimulated a search for environmental agents.*

- *Agents undergoing therapeutic trials at present include CNTF, IGF1 glutamate antagonists, branched-chain amino acids and TRH analogue.*

We are grateful to the Motor Neurone Disease Association of Great Britain for financial support for our research.

References

1. Leigh, P.N., Whitell, H., Garofalo, O. *et al.* (1991) Ubiquitin-immunoreactive intraneuronal inclusions in amyotrophic lateral sclerosis: morphology, distribution, & specificity. *Brain* **114,** 775-788

2. Whitehouse, P.J., Wamskey, J.K. Zarbin, M.A. (1983) Amyotrophic lateral sclerosis: alterations in neurotransmitter receptors. *Ann. Neurol.* **14,** 8-16

3. Hayashi, H., Fuga, M. & Satake, T. (1981) Reduced glycine receptor in spinal cord in amyotrophic lateral sclerosis. *Ann. Neurol.* **9,** 292-294

4. Plaitakis, A., Constantakakis, E. & Smith, J. (1988) The neuroexcitoxic amino acids glutamate and aspartate are altered in the spinal cord and brain in amyotrophic lateral sclerosis. *Ann. Neurol.* **24,** 446-449

5. Rothstein, J.D., Martin, L.J. & Kuncl, R.W. (1992) Decreased glutamate transport by the brain and spinal cord in amyotrophic lateral sclerosis. *N. Engl. J. Med.* **326,** 1464-1468

6. Murakami, T. (1990) Motor neuron disease: quantitative, morphological and microdensitophotometric studies on neurons of anterior horn and ventral root of cervical spinal cord with special reference to the pathogenesis. *J. Neurol. Sci.* **99,** 101-115

7. Clark, A.W., Tran, P.M., Parhad I.M. *et al.* (1990) Neuronal gene expression in amyotrophic lateral sclerosis. *Mol. Brain. Res.* **7,** 75-83

8. Virgo, L., de Belleroche, J., Rossi, M. *et al.* (1992) Characterization of the distribution of choline acetyltransferase messenger RNA in human spinal cord and its depletion in motor neurone disease. *J. Neurol. Sci.* **112,** 126-132

9. Gillberg, P.G., Aquilinius, S.M., Eckernas, S.A. *et al.* (1982) Choline acetyltransferase and substance P-like immunoreactivity in lumbar spinal cord: changes in amyotrophic lateral sclerosis. *Brain. Res.* **250**, 394-397

10. Parhad, I.M., Oishi, R. & Clark, A.W. (1992) GAP-43 gene expression is increased in anterior horn cells of amyotrophic lateral sclerosis. *Ann. Neurol.* **31**, 593-597

11. Siddique, T., Figlewicz, D.A., Pericak-Vance, M.A. *et al.* (1991) Linkage of a gene causing familial amyotrophic lateral sclerosis to chromosome 21 and evidence of genetic-locus heterogeneity. *N. Engl. J. Med.* **324**, 1383-1385

12. Weissenbach J., Gyapay G., Dib C. *et al.* (1992) A second generation linkage map of the human genome. *Nature (London)* **359**, 794-801

13. King, A., Houlden, H., Hardy, J. *et al.* (1993) Absence of linkage between chromosome 21 loci and familial amyotrophic lateral sclerosis. *J. Med. Genet.* **30**, 318.

14. Rosen, D.R., Siddique, T., Patterson, D. *et al.* (1993) Mutations in Cu/Zn superoxide dismutase gene are associated with familial amyotrophic lateral sclerosis. *Nature (London)* **362**, 59-62

15. Allen, C.N., Ross, S.M. & Spencer, P.S. (1990) Properties of the neurotoxic nonprotein amino acids, ß-N-methylamino-L-alanine (BMAA) and ß-N-oxalylamino-L-alanine (BOAA). In *Amyotrophic Lateral Sclerosis: New Advances in Toxicology and Epidemiology* (Rose, F.C. & Norris, F.H., eds.), pp. 41-48, Smith-Gordon, London

16. Yase, Y. (1990) The role of metal/mineral metabolism in the amyotrophic lateral sclerosis process. The epidemiological and environmental background. In *Amyotrophic Lateral Sclerosis: New Advances in Toxicology and Epidemiology* (Rose, F.C. & Norris, F.H., eds.), pp. 197-203, Smith-Gordon, London

17. Appel, S.H., Engelhardt, J.I., Garcia, J. *et al.* (1991) Autoimmunity and ALS: a comparison of animal models of immune-mediated motor neuron destruction and human ALS. In *Adv. Neurol.* **56**, 405-412

18. Salazar-Grueso, E.F., Routbort, M.J., Martins, J. *et al.* (1990) Polyclonal IgM anti-GM1 ganglioside antibody in patients with motor neuron disease and variants. *Ann. Neurol* **27**, 558-563

19. Sendtner, M., Schmalbruch, H., Stockli, K.A. *et al.* (1992) Ciliary neurotrophic factor prevents degeneration of motor neurons in mouse mutant progressive motor neuropathy. *Nature (London)* **358**, 502-504

20. Weiner, L.P. (1980) Possible role of androgen receptors in amyotrophic lateral sclerosis: a hypothesis. *Arch. Neurol.* **37**, 129-131

21. La Spada, A.R., Wilson, E.M., Lubahn, D.B. *et al.* (1991) Androgen receptor gene mutations in X-linked spinal and bulbar muscular atrophy. *Nature (London)* **352**, 77-79

22. Zilkha K.J. (1962) Discussion on motor neurone disease. *Proc. R. Soc. Med.* **55**, 1028-1029

23. Armon, C., Kurland, L.T., Smith, G.E. et al. (1992) Sporadic and Western Pacific amyotrophic lateral sclerosis: epidemiological implications. In *Handbook of Amyotrophic Lateral Sclerosis* (Smith, R.A. ed.), pp. 93-131, M Dekker Inc, New York

24. Olney J.W. (1986) Inciting excitotoxic cytocide among central neurons. *Adv. Exp. Med. Biol.* **203**, 631-645

25. Calne, D.B., Eisen, A., McGeer, E. *et al.* (1986) Alzheimer's disease, Parkinson's disease, and motorneurone disease: abiotropic interaction between ageing and environment? *Lancet* **ii** (8515), 1067-1070

26. Eisen, A.A., Kim, S. & Pant, B. (1992) Amyotrophic lateral sclerosis (ALS): a phylogenetic disease of the corticomotorneuron? *Muscle Nerve* **15**, 219-228

27. Appel, S.H. (1981) A unifying hypothesis for the cause of amyotrophic lateral sclerosis, Parkinsonism, and Alzheimer's disease. *Ann. Neurol.* **10**, 499-505

28. Kiernan J.A. & Hudson A.J. (1991) Changes in sizes of cortical and lower motor neurons in amyotrophic lateral sclerosis. *Brain* **114**, 843-853

29. Karpati, G., Carpenter, S. & Durham, H. (1988) Hypothesis for the pathogenesis of amyotrophic lateral sclerosis. *Rev. Neurol. (Paris)* **144**, 672-675

30. Breuer A.C., Lynn M.P., Atkinson M.B. et al. (1987) Fast axonal transport in amyotrophic lateral
     sclerosis: an intra-axonal organelle traffic analysis. *Neurology* **37**, 738-748
31. Bradley W.G. & Krasin F. (1982) DNA hypothesis of amyotrophic lateral sclerosis. In *Human
     Motor Neuron Diseases* (Rowland, L.P. ed.), pp. 493-502, Raven Press, New York
32. Tandan, R., Robison, S.H., Munzer, J.S. *et al.* (1987) Deficient DNA repair in amyotrophic lateral
     sclerosis cells. *J. Neurol. Sci.* **79**, 189-203
33. Heafield, M.T., Fearn, S., Steventon, G.B. *et al.* (1990) Plasma cysteine and sulphate levels in
     patients with motor neurone, Parkinson's and Alzheimer's disease. *Neurosci. Lett.* **110**, 216-220
34. Testa, D., Caraceni, T. & Fetoni, V. (1989) Branched-chain amino acids in amyotrophic latral scle-
     rosis. *J. Neurol.* **236**, 445-447
35. Guiloff, R.J. (1989) Use of TRH analogues in motorneurone disease. *Ann. N.Y. Acad. Sci.* **553**, 399-
     421

4

Carnitine and its role in acyl group metabolism

Rona R. Ramsay

Department of Biochemistry and Biophysics, University of California, San Francisco, and Molecular Biology Division, U.S. Department of Veterans Affairs Medical Center, San Francisco, CA 94121, U.S.A.

Importance of carnitine in metabolism[1]

The first disorder of fatty acid oxidation to be identified was a muscle carnitine palmitoyltransferase (CPT) deficiency in 1973. Since then, a variety of clinical conditions has been shown to result from defects in fatty acid metabolism. CPT deficiencies can cause fasting coma, hepatomegaly, cardiomyopathy, muscle weakness and pain, renal problems and brain dysplasia[1]. Biochemically, the build-up of abnormal CoA derivatives results in low free CoA which inhibits oxidative metabolism. The acyl groups can be excreted as carnitine derivatives, which results in low intracellular carnitine. Carnitine deficiency, either primary or secondary, can have mild to severe consequences, from muscle weakness after exercise to early death. The many disturbances of lipid metabolism accompanying such defects emphasize the central role of carnitine in acyl group transport and the importance of its modulation of the acyl-CoA/CoA ratio in the cell.

Carnitine — synthesis, disposal and cellular levels[2,3]

The history of the identification of carnitine, its chemistry, synthesis and excretion have been reviewed by Bremer[2]. L-Carnitine (β-hydroxy-α-N-trimethylaminobutyric acid, see Figure 1) is widely distributed in both animals and plants. It is synthesized from protein-bound lysine which is methylated by S-adenosylmethionine. Free trimethyl-lysine is converted to butyrobetaine in

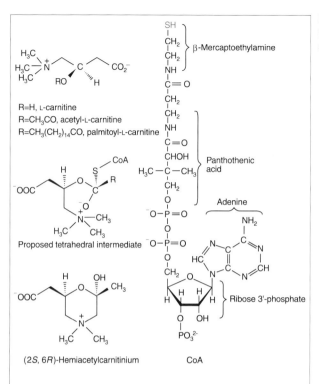

Figure 1. Substrates and inhibitors of carnitine acyltransferases

all tissues, but the final hydroxylation takes place only in liver and, to a lesser extent, in kidney and brain. Breakdown is negligible in animals and so carnitine is turned over as a result of excretion in the urine either as free carnitine or as acyl-L-carnitine. The half-life of carnitine in human liver is fairly short (11 h) compared with skeletal muscle (8 days), where carnitine is not synthesized. Plasma contains very low levels of carnitine (30-50 µM), but cell membranes contain tissue-specific, Na$^+$-dependent, carnitine-uptake systems which maintain the high levels inside the cells[6]. In heart, carnitine amounts of about 20 µmol/g dry wt. translate to about 4 mM distributed throughout the cell. In liver, the concentration is slightly lower (2 mM). The highest levels of carnitine (60 mM) are found in rat epididymal fluid, but the level is achieved by two successive transport systems, one on each side of the epididymal cells.

Carnitine and its acyl derivatives pass only slowly through membranes[2], so loss from the cell is slow ($t_{1/2}$ = 2.6 h), except in liver and kidney where carnitine is exported for use in other tissues. Within the cell there are two pools, the larger in the cytoplasm, the other in the mitochondrial matrix. The concentration of the total carnitine on each side of the mitochondrial membrane is approximately equal thanks to the action of the carnitine acyltranslocase in the mitochondrial inner membrane. The translocase can both facilitate equilibration of total carnitine, depending on the concentration gradient across the membrane, and exchange long-chain acyl-carnitine (high affinity) for free carnitine (low affinity). The latter exchange is its special role in fatty acid oxida-

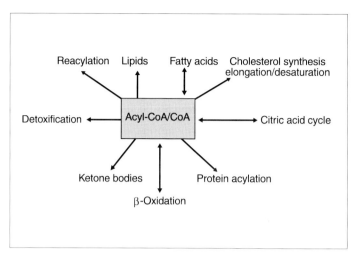

Figure 2.
Pathways using
acyl-CoA and
CoA

tion. Without the carrier, the supply of activated fatty acyl groups to the β-oxidation complex inside the mitochondrion would be too slow to support the energy needs of the cell, particularly in heart where fat is a major fuel.

The activated form of fatty acids, which is used both for β-oxidation and for lipid synthesis, is acyl-CoA (see Figure 1). Unlike carnitine, CoA is com-

$$CH_3(CH_2)_nCO_2^- + ATP + CoA \rightarrow CH_3(C_2)_nCOS\ CoA + AMP + 2P_i$$

partmented differentially, most being in the mitochondrial matrix, where acyl-CoA is important for the glycolytic pathway (pyruvate dehydrogenase), for the tricarboxylic acid cycle (α-ketoglutarate dehydrogenase) and for β-oxidation (Figure 2). In the cytosol, the concentration of CoA is very low and, indeed, most acyl-CoA groups are probably enzyme-bound for much of the time. CoA-requiring reactions in the cytosol include the synthesis of fatty acids, triacylglycerols and phospholipids (Figure 3). The much larger carnitine pool in the cytosol forms a mobile pool unifying the more localized CoA sub-pools. Cytosolic carnitine also 'buffers' the acylation state of the mitochondrial CoA by providing a means of removing the excess acetyl and longer chain acyl groups from the mitochondrial matrix (Figure 3) to maintain sufficient free CoA for optimal respiratory function. Activated acyl groups thus

$$CoA-S-C-(CH_2)_nCH_3$$
$$\underset{O}{||}$$

Thioester bond to CoA

exported form a store, ready for immediate use, which is an important transient energy source in insect flight muscle, heart and sperm[2]. The enzymes which catalyse this isoenergetic transfer of acyl groups are the carnitine acyltransferases.

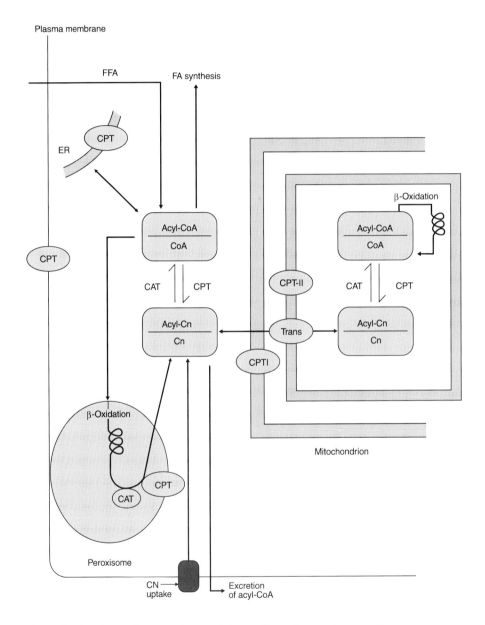

Figure 3. Carnitine, CoA and the carnitine acyltransferases in the cell
The enzymes shown are CAT (EC 2.3.1.7.) for the short-chain acyl-CoA: carnitine acyltransferases and CPT (EC 2.3.1.21) for the long-chain acyl-CoA: carnitine acyltransferases. Other abbreviations are: Cn, carnitine; ER, endoplasmic reticulum; Trans, the carnitine translocase; FA, fatty acids; FFA, free fatty acids.

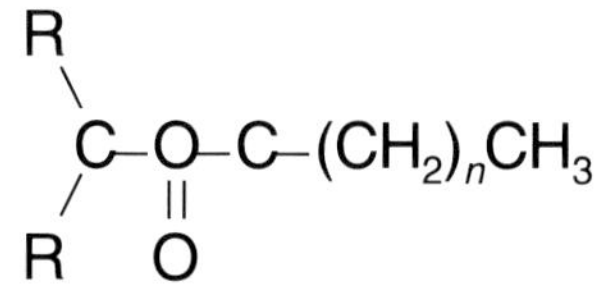

$$R$$
$$\overset{\displaystyle R}{\underset{\displaystyle R}{\diagdown}}C\text{-O-}\overset{O}{\underset{\|}{C}}\text{-}(CH_2)_n CH_3$$

Ester bond to carnitine

Carnitine acyltransferases[3,4]

The family of L-carnitine:acyl-CoA transferase enzymes catalyse the reversible equilibration of the acylation state of the CoA and carnitine pools. The over-lapping specificity of the different enzymes covers aliphatic acyl chains from C_2 to C_{24}, the β-oxidation intermediates, dicarboxylic acids (such as succinate) and a wide variety of xenobiotics (e.g. valproyl-CoA). Although all of them catalyse the same reaction (Figure 4), the properties of each isoenzyme are tuned to its location and function (Figure 3). The enzymes, their locations and some properties are listed in Table 1. For example, the CPT enzymes that use cytoplasmic substrates — those in the mitochondrial outer membrane, the per-oxisomes and the endoplasmic reticulum — all have low K_m values for the acyl-CoA and carnitine substrates, so that when even a small proportion of the CoA pool becomes acylated, the net flux will be towards the formation of acyl-L-carnitine. Inside the mitochondrion, CPT-II has a K_m for L-carnitine (1.5 mM) which is ten times higher than for CPT-I and carnitine octanoyl

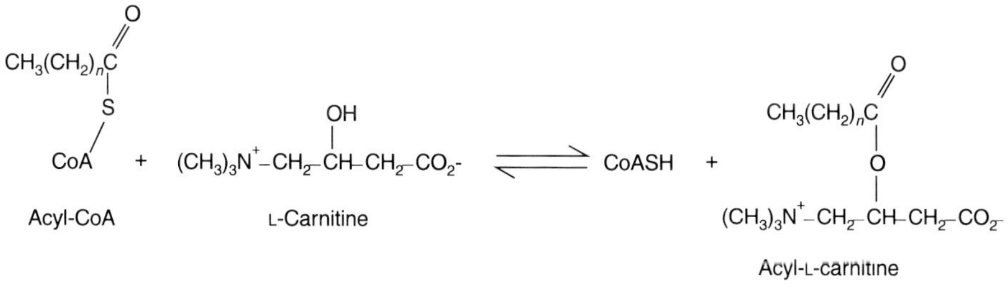

$$K_{eq.} = \frac{[\text{Acyl-L-carnitine}][\text{CoA}]}{[\text{Acyl-CoA}][\text{L-carnitine}]} \approx 1$$

$$\therefore \quad \frac{\text{Acyl-CoA}}{\text{CoA}} = \frac{\text{Acyl-L-carnitine}}{\text{L-Carnitine}}$$

Figure 4. Carnitine acyltransferases catalyse the equilibration of the acylation state of the CoA and carnitine pools

Table 1. Carnitine acyltransferases

Acronym	Name (location)	Acyl chain specificity	Malonyl-CoA sensitivity	Acyl substrate	Approx K_m (µM)				Ref.
					AcCoA	Cn	AcCn	CoA	
CAT	Carnitine acetyltransferase (mitochondria, peroxisomes, endoplastic reticulum)	C_2-C_{14}	NO	(C_2)	30	100-300	300	30	4,7
CPT-I	Outer carnitine palmitoyltransferase (mitochondrial outer membrane)	C_6-C_{18}	YES	(C_{16})	70*	100-500			8,15,16
CPT-II	Inner carnitine palmitoyltransferase (mitochondrial inner membrane)	C_8-C_{18}	NO	(C_{10})	2	1500	100	112	2,3,12
COT	Carnitine octanoyltransferase (peroxisomes)	C_6-C_{22}	YES	(C_{16})	0.6	100	31	20	9,11
CPT(ER)	Microsomal carnitine acyltransferase (endoplastic reticulum)	C_8-C_{22}	YES	(C_{16})	1-2	400	>1000		10
CPT(RBC)	Carnitine palmitoyltransferase (erythrocyte plasma membrane)	C_8-C_{22}	YES			180			7
CPT(SR)	Carnitine palmitoyltransferase (sarcoplasmic reticulum)	C_8-C_{22}	YES			500			22

*In the presence of 1.3 mg/ml bovine serum albumin.

transferase (COT) (about 0.1 mM). Since the L-carnitine concentration is about 2 mM (in both the matrix and the cytoplasm) and acyl-CoA is removed by β-oxidation, the net flux in the matrix will normally favour the reverse reaction to produce acyl-CoA for β-oxidation (Figure 3).

Analogues of L-carnitine have been used to explore the differences in the carnitine-binding sites. For example, L-aminocarnitine is a substrate for the enzymes using cytoplasmic substrates, but not for CPT-II. In contrast, the

$$
\underset{\text{Aminocarnitine}}{(CH_3)_3 \overset{+}{N}-CH_2-\overset{\overset{\displaystyle NH_2}{|}}{C}-CH_2CO_2^-}
$$

CoA site is remarkably similar in all the enzymes (K_s = 20-30 μM[4]). The acyl group binding site is the one which varies most and which defines the specificity of substrate binding. CAT shows good activity with C_2 and C_4 substrates, but no activity at all with palmitoyl-CoA. However, palmitoyl-CoA is a potent inhibitor no matter which substrate is varied, suggesting that the CoA moiety binds normally, but the large fatty acyl group blocks the access to the carnitine site. The long-chain acyltransferases (see Table 1) all have a broad spectrum of activity from C_8 to C_{24}, some better at the shorter end (e.g. mouse liver COT) and some better at the longer end (e.g. CPT$_{RBC}$, which has good activity with C_{18}-C_{24}, a range which includes the fatty acids common in the erythrocyte membrane[7]). The acyl-binding pocket contributes significantly to binding, as seen for COT where the K_s drops from 20 μM for CoA to 0.5 μM for decanoyl-CoA.

To develop effective inhibitors of the transferase reaction, Gandour[4] synthesized conformationally rigid analogues of acyl-L-carnitine which mimic the proposed transition state tetrahedral intermediate (Figure 1). One of these, hemipalmitoyl-carnitinium, is a potent inhibitor of CPT-II (K_i = 0.2 μM).

The kinetic mechanism has been determined for three of the isoenzymes. Carnitine acetyl transferase (CAT) and COT follow random-order rapid equilibrium kinetics, where neither substrate affects the binding of the other, but CPT-II follows an ordered mechanism in which CoA or acyl-CoA must bind first. Thus, the acyl-CoA/CoA ratio in the mitochondrial matrix will strongly influence the direction of the reaction.

The reversible equilibrium nature of the reaction ensures a mass action response to the acyl-CoA pool (Figure 4). However, superimposed on that immediate response is a regulation by malonyl-CoA of the CPT enzymes which use cytoplasmic substrates (Figure 3). The regulation by malonyl-CoA, the building block for fatty acid synthesis, integrates fatty acid oxidation and

$$CO_2^-$$
$$|$$
$$CH_2$$
$$|$$
$$O = C - SCoA$$

Malonyl-CoA

fatty acid synthesis to prevent futile (energy-expensive) cycling of fatty acids between oxidative and synthetic pathways[8]. The inhibition of CPT-I by malonyl-CoA regulates the flow of fatty acids into the mitochondrion to decrease β-oxidation and, consequently, the production of ketone bodies in liver[8]. The acyltransferases in the peroxisomes[9] and the endoplasmic reticulum[10] are also regulated by malonyl-CoA. In starvation, malonyl-CoA levels drop and the sensitivity of CPT to malonyl-CoA inhibition decreases, so that the acyltransferase activity is maximized and more β-oxidation is possible, i.e. the cell switches to a fat fuel source. CPT-I is the most important enzyme in this regulation. It provides 65% of the cytoplasmic long-chain acyltransferase activity in the liver. It is inhibited by micromolar concentrations of malonyl-CoA and the inhibition by malonyl-CoA can be overcome by high concentrations of palmitoyl-CoA[10]. The levels of the CPT enzymes also change, and increases in CPT activity, immunoreactive protein and mRNA levels have been reported in starved, diabetic or hypothyroid animals. High-fat diets and drugs such as clofibrate, a peroxisome proliferator, also increase CPT levels. Glucagon increases CPT activity, but insulin decreases it. This hormonal control emphasizes the central role of CPT in the regulation of fatty acid metabolism. That inhibition of CPT-I prevents ketone body formation and helps glucose disposal, supports the concept of the importance of the regulation of fatty acid metabolism in diabetes[11].

The molecular biology of these enzymes still has a long way to go. Only COT[12] and CPT-II[13,14] have been cloned. (Note added in proof: CPT-I has now been cloned[23].)When conservative substitutions are taken into account, the cDNA sequences suggest that there is about 50% similarity between their amino acid sequences and that they share the CoA-binding site motif common to most CoA-dependent proteins and some conserved residues which may be relevant to carnitine binding. The levels of transcription change in parallel with the activity measured and the regulation of expression is now being explored.

The evidence surrounding the identity of CPT-I, the key enzyme in the regulation of fatty acid oxidation is complex. (i) Only one active protein (68 kDa, i.e. CPT-II) has been isolated from purified mitochondria. However, CPT-I activity is easily destroyed by the detergents used to solubilize membrane proteins. (ii) Malonyl-CoA-sensitive CPT-I is found on the outer membrane of the mitochondrion[15], but it has been suggested that some inner membrane proteins could be carried over in contact sites during the isolation of the outer membranes. (iii) Measurement of the molecular mass of the CPT activities in rat liver mitochondria *in situ* by target inactivation analysis identified a 70 kDa target as CPT-II and an 83 kDa target as CPT-I. Malonyl-CoA binding was associated with a different target, one of 60 kDa. (iv) All malonyl-CoA-sensitive enzymes are irreversibly inactivated by tetradecylglycidyl-

$$CH_3(CH_2)_{13}-\underset{\underset{H_2C-O}{\diagup\diagdown}}{C}-CO-SCoA$$

2-Tetradecylglycidyl-CoA

CoA. The modified protein isolated from heart mitochondria was 90 kDa and from skeletal muscle was 86 kDa indicating that there are tissue-specific isoenzymes. (v) Antibody studies indicate that the tetradecylglycidyl-CoA-inactivated protein is indeed responsible for CPT-I activity and is different from CPT-II[16]. (vi) A malonyl-CoA-binding protein (86 kDa) has been isolated which confers malonyl-CoA sensitivity on the purified 68 kDa protein (CPT-II)[17]. (vii) The only gene product as yet linked to mitochondria is the 68 kDa CPT-II.

CPT-I is now known to be a separate gene product[23] but much more information is required to understand the regulation of CPT-I. Because CPT-I regulates the flux of fatty acids into the β-oxidation pathway, it is a central target for the development of drugs, useful in medical and dietary pharmacology.

Carnitine translocase[2,3]

The oxidation of fatty acids was shown to be carnitine and CPT dependent in 1955, but it was 20 years later before the mechanism by which carnitine facilitates the transfer of acyl groups through the membrane was discovered. Carnitine exchange across the mitochondrial inner membrane is catalysed by a carrier (see Figure 3), which works equally well with all acyl-carnitines, but the affinity increases with chain length, so that palmitoyl-L-carnitine is effective at concentrations two orders of magnitude lower than is L-carnitine. The carrier has been isolated and reconstituted into vesicles[18]. The kinetics of exchange studied in these vesicles indicate a Ping-Pong mechanism which means that the carrier binds only one ligand at a time and re-orientates from one side of the membrane to the other. The unloaded carrier can re-orient slowly, providing a mechanism for adjusting the level of carnitine in the matrix in response to a stimulus such as starvation, which increases the total carnitine pool. The carrier may normally exist in a functional complex with CPT-II and the β-oxidation enzymes for the efficient oxidation of fatty acids.

Other carnitine-dependent processes

Product removal in peroxisomes

Before leaving the topic of fatty acid oxidation, β-oxidation in peroxisomes must be considered. Peroxisomal oxidation accounts for very little of the β-oxidation in heart, but up to 50% of the long-chain fatty acid metabolism in mouse liver, especially for fatty acids longer than C_{18}. The oxidation does not require carnitine, but carnitine may be important for the transfer of the chain-shortened, activated fatty acids from the peroxisomes to the mitochondria for complete oxidation. The specificity of the COT present in the peroxisomes may play a role in determining the point of chain termination. The COT from mouse liver has highest activity with C_8 chain length substrates (octanoyl-CoA), whereas rat liver COT is most active with C_{10} and C_{14}, and beef liver COT with C_{16} acyl-CoA. In this respect, carnitine fulfils a transport or chaperone role. CAT is also present in peroxisomes, presumably to facilitate the removal of the acetyl groups produced in β-oxidation from the limited pool of CoA available in the peroxisomes.

Excretion of acyl groups

Carnitine is not metabolized to any great extent in animals, but is excreted in the urine. It is a polar compound capable of forming esters with a wide variety of carboxylic acids, so it is not surprising that it is a vehicle for the excretion of many excess acyl groups. Pathological build up of the intermediates of fatty acid oxidation which results from deficiencies in acyl-CoA dehydrogenases is alleviated by carnitine. Free carnitine is re-absorbed by the kidneys, but secondary carnitine deficiencies are often found when large amounts of acyl-carnitines are excreted. Xenobiotics which form acyl-CoA thioesters can also be transferred to carnitine for excretion. Valproate, an anti-convulsant, is metabolized to valproyl-CoA, sequestering CoA, so carnitine is often administered as an adjuvant to promote its excretion.

Carnitine in heart disease[1]

Severe carnitine deficiencies are associated with cardiomyopathy. Therapy with carnitine in these cases, and in patients with defects in the plasma membrane uptake system usually results in progressive improvement. Since free fatty acids are the preferred metabolic fuel for energy production in the heart, alterations in myocardial levels of carnitine can alter that metabolism and adversely affect contractile performance. Although congestive heart failure patients rarely have low myocardial carnitine levels, long-chain acyl-carnitines can build up and these can inhibit mitochondrial enzymes, such as the ADP transporter, in addition to direct physical effects on the membrane. Increased myocardial carnitine has a protective effect in the ischaemic heart, as do CPT-I inhibitors. When used during re-perfusion of ischaemic hearts, these agents stimulate glucose oxidation presumably by reducing the acyl-CoA/CoA ratio in the mitochondria.

Carnitine also protects against adriamycin-induced cardiotoxicity. Adriamycin is an antibiotic used as a chemotherapeutic agent against a variety of cancers. The toxic side-effects result from inhibition of mitochondrial energy production at the level of cytochrome oxidase, perhaps owing to binding of the drug to cardiolipin. Carnitine may protect by blocking that binding.

Direct effects of acyl-carnitines

Long-chain acyl-carnitines have a polar head-group and a hydrophobic tail (Figure 1), and can form micelles, disrupt membranes and bind to many proteins. Aside from detergent-like effects, such as changing membrane fluidity, binding at membrane or protein hydrophobic sites can alter the properties of the binding target. The protective effect of carnitine against adriamycin (see above) is a good example. Because acyl-carnitines build up in the heart during anoxia, their deleterious effects have been studied. For example, acyl-carnitines (either D- or L-) may modulate Ca^{2+}-channel function. That acyl-D-carnitine is effective is a clear indication that the mechanism is a direct one, because only L-carnitines are substrates for the acyltransferases. Peroxisome proliferation can be induced by many agents, such as plasticizers, clofibrate and very-long-chain fatty acids. A potential role for acyl-carnitines in the induction has not yet been established. However, inhibitors of COT prevent the proliferation, suggesting that either the formation of acyl-carnitine is important or that the resulting build-up of acyl-CoA in the peroxisomes inhibits the triggering of proliferation.

Cell-cell interactions

Recently, a novel effect of carnitine was reported[19]. Carnitine (either D- or L-) inhibits the specific cell-adhesion processes of the slime mould, *Dictyostelium discoideum*. This process involves electrostatic interaction of cell surface proteins. It was postulated that the charged ends of carnitine bind to the aspartate and lysine groups increasing hydration around the site and preventing the approach of the other cell.

The regulatory role of carnitine and the carnitine acyltransferases[5]

Carnitine acyltransferases, by catalysing the transfer of short-chain acyl groups to carnitine, are important in the regeneration of free CoA. For example, the addition of carnitine stimulates the oxidation of pyruvate by isolated mitochondria and acetyl-carnitine builds up in the medium. Heart mitochondria have a high capacity for this export of acetyl groups from the matrix CoA pool to the larger external carnitine one. Conversely, the flux may be reversed to import acetyl groups if energy demands reduce the acetyl-CoA/CoA ratio too far[3]. Liver mitochondria contain less CAT and less carnitine, so that the export of the acetyl groups may be rate-limiting. Indeed, the equilibrium of the mitochondrial acetyl-CoA/CoA and cytoplasmic acetyl-L-carnitine/carnitine which would be predicted on the basis of the properties of CAT and the translocase, is not always found in liver. The consequences of the lack of free CoA are severe, such as in Reye's syndrome, where as much as 90% of the CoA pool in the liver is acylated. Although not yet established, it is possible that the improvement of brain function in elderly patients on long-term acetyl-L-carnitine therapy is owing to modulation of the acetyl-CoA/CoA ratio in brain cells and the consequent improvement in energy production[20].

CPT-II plays a similar role for long-chain acyl groups within the mitochondrion. In patients with acyl-CoA dehydrogenase deficiencies, the intermediates of β-oxidation which build up are exported as carnitine derivatives. In peroxisomes, CAT and COT remove acetyl and medium-chain acyl products from the peroxisomal CoA pool, again maintaining the local acyl-CoA/CoA ratio within the limits favourable for the continuing function of the organelle. Microsomes, where long-chain fatty acyl-CoA is the source of acyl groups for lipid synthesis and for re-acylation of lysophospholipids, also contain CPT. Here, too, a buffering role is probable, but has not yet been studied in detail.

The system which drew attention to the role of carnitine in modulating the acyl-CoA pool was the re-acylation repair process for phospholipids in erythrocytes[7]. There is no fatty acid oxidation pathway in erythrocytes, so this re-acylation process is the main use of fatty acids. Normally, free fatty acids are activated to acyl-CoA in an ATP-dependent process to provide the substrate for the lysophospholipid acyl-CoA transferases. CPT may affect the acylation process by modulating the size of the acyl-CoA pool. The long-chain acyl-carnitine formed represents an ATP-independent acyl reservoir for the re-acylation. Differences in flux into (from acyl-CoA synthetase) and out of (via lysophospholipid acyl-CoA transferase) the acyl-CoA pool could alter the acyl-CoA/CoA ratio, but the mass-action response of CPT will suppress the fluctuations. The concept of control of acyl-trafficking by CPT may also be important in other cells where re-acylation is the mechanism of remodelling the phospholipids to change membrane fluidity and of the generation of bioactive phospholipids. The CPT in erythrocytes has an extended specificity range with moderate activity with the very-long-chain fatty acids commonly found in the erythrocyte membrane (e.g. arachidonate, $C_{20:4}$). Inhibition of CPT with tetradecylglycidic acid completely inhibited the incorporation of radiolabelled fatty acids from acyl-carnitine into membrane fatty acids, and decreased the incorporation from free fatty acid when the substrate was oleate, but not when it was palmitate. Oleoyl-CoA is synthesized much faster than it is incorporated into phospholipid, so that the CoA pool could become over-acylated. The rate of generation of oleoyl-CoA by the synthetase is 1.3 times that for palmitoyl-CoA. For the incorporation into phosphatidylcholine the preference for oleate is 2.7 times. This imbalance of rates would result in the accumulation of acyl-CoA, particularly palmitoyl-CoA which is used less in re-acylation reactions. However, the preferred substrate for CPT is palmitoyl-CoA and the rate with oleoyl-CoA is only 0.6 of that. Thus, CPT catalyses the removal of the acyl group less favoured for incorporation and modulates the optimal acyl-CoA/CoA ratio for the physiological expression of fatty acid turnover in membrane phospholipids.

Future directions

The versatile function of carnitine and the carnitine acyltransferases in intracellular metabolism is best summarized as an interconnected capacitor system buffering fluctuations in the acyl-CoA/CoA ratio in multiple pools. The import of activated fatty acids into mitochondria for β-oxidation is a gated process where regulation of CPT serves to integrate fatty acid breakdown and synthesis, and as a point of hormonal control. The export of short-chain acyl groups from both mitochondria and peroxisomes is a direct response to changes in the acylation state of the CoA pool. Similarly, excretion of undesirable acyl groups is a response to accumulation in the CoA pool.

The future of carnitine research lies in exploring the ramifications of the modulation of the acyl-CoA/CoA pools. This process has been initiated using specific inhibitors, for example of CAT, to explore the resulting deviations from the normal metabolic pattern[21]. Etomoxiryl-CoA which inhibits CPT-I, and is already in clinical use in the parent free fatty acid form, can be used similarly both in cell culture and whole animals to explore the consequences of inhibition of CPT-I beyond the restriction of β-oxidation.

Etomoxiryl-CoA

Questions about the carnitine acyltransferases themselves also remain to be explored. Our understanding of their genetic origin, control of their levels, the effects of the membrane environment on their properties and the mechanism of malonyl-CoA regulation remains incomplete.

My research is supported by the National Institutes of Health (HL-16251 and DK-41572) and by the National Science Foundation (DMB 9020015).

References

References 1-5 are recommended for further reading. To keep the list short, recent references which refer back to the pioneer work are given where possible.

1.   Carter, A.L. (ed.) (1991) *Current Concepts in Carnitine Research*

2.   Bremer, J. (1983) Carnitine — metabolism and functions. *Physiol. Rev.* **63**, 1420-1479

3.   Bieber, L.L. (1988) Carnitine. *Annu. Rev. Biochem.* **57**, 261-283

4.   Colucci, W.J. & Gandour, R.D. (1988) Carnitine acetyltransferase: a review of its biology, enzymology, and bioorganic chemistry. *Bioorg. Chem.* **16**, 307-334

5.   Ramsay, R.R. & Arduini, A. (1993) The carnitine acyltransferases: their effect on acyl-CoA pools and the consequences for metabolism. *Arch. Biochem. Biophys.* **302**, 307-314.

6.   Stanley, C.A. (1992) Plasma and mitochondrial membrane carnitine transport defects, in *New Developments in Fatty Acid Oxidation* (Coates, P.M. & Tanaka, K., eds.), pp. 289- 300, Wiley-Liss, Inc., New York.

7.   Arduino, A., Mancinelli, G., Radatti, G.L., Dottori, S., Molajoni, F. & Ramsay, R.R. (1992) Role of carnitine and carnitine palmitoyltransferase as integral components of the pathway for membrane phospholipid fatty acid turnover in intact human erythrocytes. *J. Biol. Chem.* **267**, 12673-12681

8.   McGarry, J.D., Leatherman, G.F. & Foster, D.W. (1978) Carnitine palmitoyltransferase I. The site of inhibition of hepatic fatty acid oxidation by malonyl-CoA. *J. Biol. Chem.* **253**, 4128-4136

9.   Derrick, J.P. & Ramsay, R.R. (1989) L-Carnitine acyltransferase in intact peroxisomes is inhibited by malonyl-CoA. *Biochem. J.* **262**, 801-806

10.   Lilly, K., Bugaisky, G.E., Umeda, P.K. & Bieber, L.L. (1990) The medium-chain carnitine acyltransferase activity associated with rat liver microsomes is malonyl-CoA sensitive. *Arch. Biochem. Biophys.* **280**, 167-174

11.   McGarry, J.D. (1992) What if Minkowski had been ageusic? An alternative angle on diabetes. *Science* **258**, 766-770

12.   Chatterjee, B., Song, C.S., Kim, J.M. & Roy, A.K. (1988) Cloning, sequencing and regulation of rat liver carnitine octanoyltransferase: transcriptional stimulation of the enzyme during peroxisomal proliferation. *Biochemistry* **27**, 9000-9006

13.   Woeltje, K.F., Esser, V., Weis, B.C., Sen, A., Cos, W.F., McPhaul, M.J., Slaughter, C.A., Foster, D.W. & McGarry, J.D. (1990) Cloning, sequencing, and expression of a cDNA encoding rat liver mitochondrial carnitine palmitoyltransferase II. *J. Biol. Chem.* **265**, 10720-10725

14.   Finocchiaro, G., Taroni, F., Rocchi, M., Martin, A.L., Colombo, I., Tarelli, G.T. and DiDonato, S. (1991) cDNA cloning, sequence analysis and chromosomal localization of the gene for human carnitine palmitoyltransferase. *Proc. Natl. Acad. Sci. U.S.A.* **88**, 661-665

15.   Murthy, M.S.R. & Pande, S.V. (1990) Characterization of a solubilized malonyl-CoA-insensitive carnitine palmitoyltransferase of the inner membrane. *Biochem. J.* **268**, 599-604

16.   Kolodziej, M.P., Crilly, P.J., Corstorphine, C.G. & Zammit, V.A. (1992) Development and characterization of a polyclonal antibody against rat liver mitochondrial overt carnitine palmitoyltransferase (CPT-I). *Biochem. J.* **282**, 415-421

17.   Chung, C.H., Woldegiorgis, G., Dai, G., Shrage, E. & Bieber, L.L. (1992) Conferral of malonyl-CoA sensitivity to purified rat heart mitochondrial carnitine palmitoyltransferase. *Biochemistry* **31**, 9777-9783

18.   Indiveri, C., Tonazzi, A., Prezioso, G. & Palmieri, F. (1991) Kinetic characterization of the reconstituted carnitine carrier for rat liver mitochondria. *Biochim. Biophys. Acta* **1065**, 231-238

19.   Siu, C.-H., Brar, P. & Fritz, I.B. (1992) Inhibition of cell-cell adhesion and morphogenesis of *Dictyostelium* by carnitine. *J. Cell. Physiol.* **152**, 157-165

20.   Carta, A. & Calvani, M. (1991) Acetyl-L-carnitine: a drug able to slow the progress of Alzheimer's disease. *Ann. N.Y. Acad. Sci.* **640**, 228-232

21.   Brass, E.P., Gandour, R.D. & Griffith, O.W. (1991) Effect of carnitine acyltransferase inhibition on rat hepatocyte metabolism. *Biochim. Biophys. Acta* **1095**, 17-22

22. McMillin, J.B., Hudson, E.K. & Van Winkle, W.B. (1992) Evidence for malonyl-CoA sensitive carnitine acyl-CoA transferase activity in the saroplasmic reticulum of canine heart. *J. Mol. Cell. Cardiol.* **24,** 259-268.

23. Esser, V., Britton, C.H., Weiss, B.C., Foster, D.W. and McGarry, J.D. (1993) Cloning, sequencing and expression of a cDNA encoding rat liver carnitine palmitoyltransferase. *J. Biol. Chem.* **268,** 5817-5822.

5

Folate/vitamin B$_{12}$ inter-relationships

John Scott and Donald Weir

Department of Biochemistry and Department of Clinical Medicine, Trinity College Dublin, Dublin 2, Republic of Ireland

Folate chemistry

Folate co-factors consist of a pteridine ring substituted in the 2- and 4-positions with an amino and an hydroxy group respectively (Figure 1). The oxidized form, folic acid, does not occur in nature and arises in biological material owing to the chemical oxidation of reduced natural forms. The vitamin was first isolated in this form and is the compound used therapeutically. It is converted to the biologically active tetrahydrofolate by the cell (Figure 2). The intracellular forms of this vitamin, of which there are 10, are always conjugated to a γ-polyglutamyl chain varying in length from 3 to 11 residues depending upon the species. The generic name for the whole range is folate[1].

Folate biochemistry

The folates are involved in so-called carbon one metabolism[1]. They function by being enzymically converted into folate co-factor forms which have a carbon group attached[2]. Other enzymes use these carbon groups in important biosynthetic reactions (Figure 2). Tetrahydrofolate can be converted to 5,10-methylenetetrahydrofolate with the methylene group being donated by one of the three carbons of the amino acid serine. The breakdown of the amino acid histidine, and the cleavage of glycine or formate, can also be used as sources of carbon one units to make this and two other carbon-one-substituted forms of tetrahydrofolate, namely 5,10-methenyl- and 10-formyltetrahydrofolate. Thus tetrahydrofolate is converted into a range of carbon-one-substituted co-factors

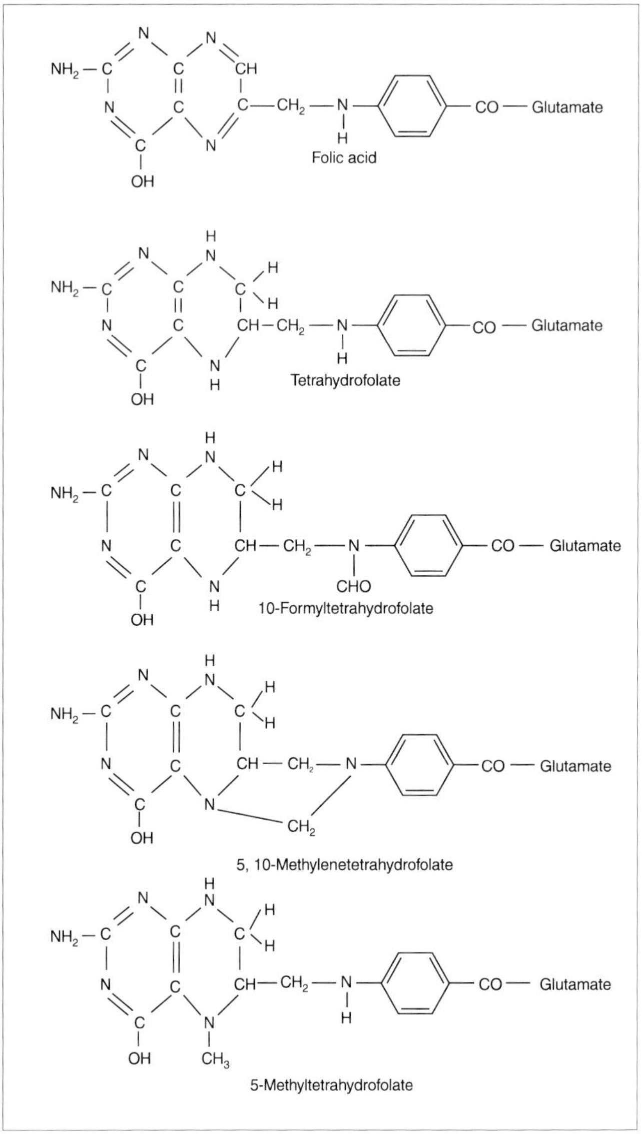

Figure 1. Formula of folic acid and tetrahydrofolic acid

which are themselves enzymically interconvertable. 10-Formyl-tetrahydrofolate acts as the co-factor for two different enzymes involved in the biosynthesis of the carbons 2 and 8 of the purine ring (Figure 2). 5,10-Methylenetetrahydrofolate is the co-factor for thymidylate synthatase, which converts the pyrimidine base uracil to the corresponding uracil base. These reactions shaded red in Figure 2 are essential for DNA and RNA biosynthesis. 5,10-Methylenetetrahydrofolate can also be reduced to 5-methyltetrahydrofolate which can then be used to convert homocysteine to methionine by means of the vitamin B$_{12}$-dependent enzyme, methionine synthase. Methionine formed in this way, or from exogenous sources, can be activated with ATP to

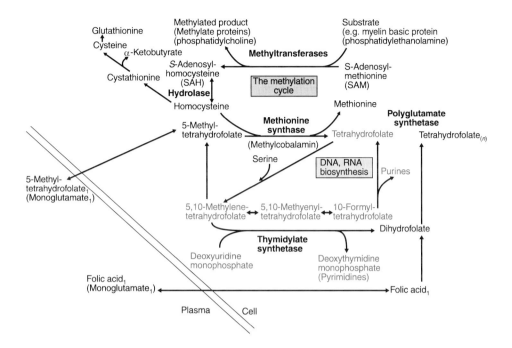

Figure 2. Pathways of folate metabolism

produce *S*-adenosylmethionine (SAM or AdoMet) in all mammalian cells. Methyltransferase enzymes use SAM as a source of methyl groups to methylate a wide range of compounds, ranging from proteins, DNA, and phospholipids, to small molecules, such as DOPA (3-hydroxy-L-tyrosine). When SAM donates a methyl group it produces *S*-adenosylhomocysteine (SAH or AdoHcy), which in turn can be enzymically hydrolysed to homocysteine. Homocysteine can then be re-cycled back to methionine, the new methyl group being provided by 5-methyltetrahydrofolate, or, alternatively, in the liver by betaine, a degradation product of choline. This methylation cycle (shown in black in Figure 2) both utilizes and provides methyl groups for cellular methyltransferases. Accordingly, folate co-factors transfer carbon groups for purine and pyrimidine biosynthesis, or as methyl groups in a wide range of cellular components.

Quantitatively, the carbon 3 of serine is the most important source of these carbon one units. All cells, particularly actively dividing cells, can make serine in three enzymic steps from glyceraldehyde 3-phosphate, which is readily available from glucose via the glycolytic pathway. Alternatively it has been suggested that formate produced by oxidation in mitochondria may also be an important source of carbon one units.

To be retained by the cell, folates taken up from the plasma must be converted into a polyglutamate. The enzyme which adds these additional glutamates will do so to any folate co-factor form, except 5-methyltetrahydrofolate, the plasma form. Thus for plasma folate to be so conjugated it must have its methyl group removed once it enters the cell and this can only be done by the vitamin B$_{12}$-dependent enzyme methionine synthase. As discussed later, one of the consequences of vitamin B$_{12}$ deficiency is decreased retention of folate by cells.

Folate deficiency

Since folate co-factors are essential for both purine and pyrimidine biosynthesis, their deficiency compromises the cells' ability to synthesize DNA and RNA and thus to divide. This reduced rate of cell division ultimately affects all cells, but initially affects rapidly dividing cells, such as those of the bone marrow or the intestinal mucosa[3]. Thus folate deficiency produces a macrocytic anaemia. The red cell precursors before they leave the bone marrow are megaloblastic and, in contrast to the usual normoblastic precursors, they have a large diffuse nucleus, which results from the slower rate of cell division. The synthesis of other cells derived from bone marrow is also impaired, resulting in a reduced number of platelets and granulocytes. The effect on other cell lines will depend upon their turnover, their requirements and the local availability of folate. Perhaps because of its ability to concentrate and conserve folate, only severe prolonged folate deficiency appears to affect the nervous system. Dietary folate deficiency, while not as common as iron deficiency, can occur in the elderly and during pregnancy. The second trimester of pregnancy puts high demands on the mother's folate stores, owing to the increased requirements for DNA synthesis and growth, which in turn causes an enhanced rate of catabolism of the vitamin during this period[4]. Many pregnant women have overt signs of folate deficiency by the end of their pregnancy if they have poor folate status to begin with and are not given prophylactic folic acid to meet the extra demand. (This folate deficiency of late pregnancy is not to be confused with the use of folic acid to prevent neural tube defects. Folic acid supplements given to women before conception and during the first trimester have recently been shown to reduce the prevalence of spina bifida and anencephalus by over two-thirds)[5]. Folate deficiency also occurs in alcoholics, patients on anti-convulsant drugs and in intestinal malabsorption states (coeliac disease and tropical sprue)[3].

Vitamin B$_{12}$ chemistry

This vitamin consists of a cobalt-substituted porphyrin ring with complex side-chains. The general name for the vitamin is cobalamin[6]. The metabolically important cobalamins differ from each other in having either a methyl group or a 5′-deoxyadenosyl group attached to this cobalt (Figure 3).

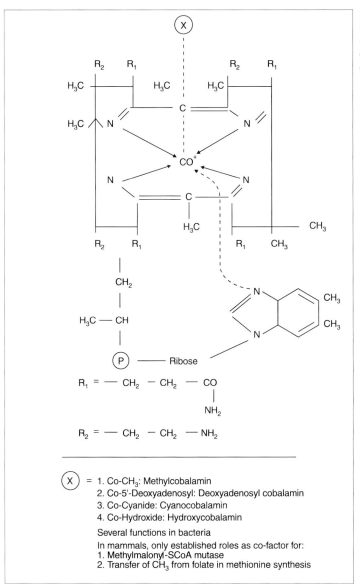

Figure 3. Formula of methyl- and deoxyadenosylcobalamin

X = 1. Co-CH₃: Methylcobalamin
2. Co-5'-Deoxyadenosyl: Deoxyadenosyl cobalamin
3. Co-Cyanide: Cyanocobalamin
4. Co-Hydroxide: Hydroxycobalamin

Several functions in bacteria

In mammals, only established roles as co-factor for:
1. Methylmalonyl-SCoA mutase
2. Transfer of CH₃ from folate in methionine synthesis

Vitamin B₁₂ biochemistry

Cyanocobalamin and hydroxy cobalamin, both of which are used therapeutically, are converted into the two active forms *in vivo*[6]. 5′-Deoxyadenosylcobalamin, one of the two active forms, is a prosthetic group for a wide range of enzymes in micro-organisms, but only one such enzyme, methylmalonyl-CoA mutase, occurs in mammalian cells. Its function is to convert methylmalonyl-CoA, a degradation product of propionate, to succinyl-CoA. In most mammals, propionate arises from the degradation of certain amino acids and after β-oxidation of odd-chain fatty acids. In ruminants it is produced by the

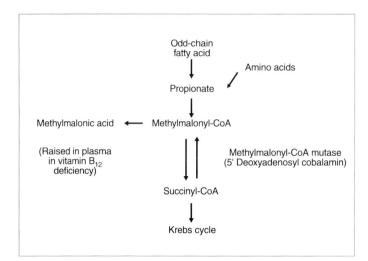

Figure 4. Pathway of propionate metabolism to methylmalonate and succinyl-CoA

microflora of the rumen and is a principal source of energy. Such propionate is thus channelled via methylmalonly-CoA to succinyl-CoA, into the Krebs cycle (Figure 4). The other active form, methylcobalamin, along with 5-methyltetrahydrofolate, is a co-factor for the enzyme methionine synthase, a key enzyme in the methylation cycle (Figure 2). As mentioned above, methionine synthase is also essential for the demethylation and thus retention of circulating 5-methyltetrahydrofolate.

Vitamin B$_{12}$ deficiency

There is a considerable excess over requirements of vitamin B$_{12}$ in most diets. The vitamin is synthesized originally by micro-organisms and enters the food chain only in food of animal origin. Thus vegetarian diets, particularly strict vegetarian, so-called vegan diets, do not contain adequate vitamin B$_{12}$. Vegans in developing countries may derive sufficient cobalamins from bacterial contamination of their diet. In developed societies vitamin B$_{12}$ deficiency is usually owing to malabsorption. Vitamin B$_{12}$ is a large molecule and unlike the other water-soluble vitamins, less than 3% is absorbed by diffusion. Absorption instead relies on the following[7]. The acidic conditions in the stomach, together with the start of protein digestion, cause vitamin B$_{12}$ to be released from food. Glycoproteins called R-binders, which have arisen in the salivary and gastric secretions, have a high affinity for all forms of vitamin B$_{12}$ at acidic pH. Simultaneously, a different glycoprotein called intrinsic factor is secreted by the parietal cells of the stomach. This mixture passes into the upper small intestine where partial digestion of the R-binder by the pancreatic digestive enzymes releases its vitamin B$_{12}$. Intrinsic factor has a higher affinity and specificity for the active forms of dietary vitamin B$_{12}$ at a neutral pH than the R-binder; accordingly, it binds the vitamin and carries it to the terminal ileum. The complex is then phagocytosed by specific calcium-dependent receptors.

Vitamin B_{12} is thus protected during its transit through the gut and only the active form of the vitamin is selected for absorption from a range of naturally occurring analogues. This process, however, can be interfered with in different ways. The secretion of intrinsic factor by the stomach may be affected. The commonest cause of reduced intrinsic factor secretion is the autoimmune disease, pernicious anaemia, which affects the gastric mucosa. In this condition the parietal cells are destroyed by an autoimmune process producing atrophic gastritis. As with other autoimmune conditions antibodies are produced, in this case against the patient's own parietal cells, and against whatever intrinsic factor is secreted. These antibodies both 'block' the complexing of intrinsic factor with cobalamin, and 'bind' to the complex preventing uptake by the ileal receptor. In the past, intrinsic factor deficiency was also produced by gastric operations for peptic ulcer disease. Nowadays, drugs which inhibit the secretion of acid by the parietal cell are used and these in turn have the potential to interfere with vitamin B_{12} absorption.

Intestinal diseases also affect vitamin B_{12} absorption. The parasite *Diphylobothrium latum* competes for the intrinsic factor vitamin B_{12} complex in the lumen of the bowel. In chronic pancreatitis, 'R' protein is not destroyed by pancreatic enzymes, which upsets the balance between the 'R' protein and intrinsic factor avidity to bind cobalamin. Crohn's disease may directly affect the ileal mucosa, or may be treated by surgical ileal resection, both of which interfere with the ileal absorption of the intrinsic factor vitamin B_{12} complex.

Vitamin B_{12} deficiency produces two very distinct clinical sequelae[3]. A megaloblastic anaemia identical to that seen in folate deficiency and a neurological condition called sub-acute combined degeneration (SCD). Vitamin B_{12} deficiency produces the megaloblastic anaemia by interfering with folate metabolism in a way that inhibits cell division and is accordingly synonomous with folate deficiency. The most widely accepted explanation for this phenomenon is called the 'methyl-trap hypothesis'[8]. It suggests that 5-methyltetrahydrofolate formed in vitamin B_{12}-deficient cells becomes metabolically trapped in this form and is thus unavailable to participate in purine and pyrimidine biosynthesis. This hypothesis requires that: (i) In the absence of vitamin B_{12} and methionine synthase, 5-methyltetrahydrofolate cannot be converted to tetrahydrofolate. While there is a vitamin B_{12}-independent version of this enzyme in bacteria, it appears that such an enzyme does not exist in mammalian cells. (ii) *In vivo* the reductase enzyme that converts 5,10-methylenetetrahydrofolate to 5-methyltetrahydrofolate cannot catalyse the reverse reaction in significant amounts. While this reaction can be made to reverse *in vitro*, and is the way that the enzyme's activity is measured, it requires a very strong electron acceptor for this to happen. Studies *in vitro* have shown that the reductase can catalyse the reverse reaction with the quinone of a different cofactor, namely dihydrobiopterin, and this may also happen *in vivo*[9]. So the reaction may not be as irreversible as is claimed by advocates of the methyl-trap hypothesis. (iii) The cell will pile up its vital folate co-factors in a single

and, because of lack of methionine synthase, unusable form, namely 5-methyl-
tetrahydrofolate.

Support for the methyl-trap hypothesis comes from the finding by
Kutzback & Stokstad[10] that the activity of the methylene reductase enzyme is
controlled *in vitro* and presumably *in vivo* by allosteric inhibition by SAM
(Figure 2). The purpose of forming 5-methyltetrahydrofolate is clearly to
channel carbon one units to re-methylate the homocysteine formed as a result
of the utilization of SAM for methyltransferase reactions. This re-synthesizes
methionine and thus maintains the levels of SAM (Figure 2). A reduction in
SAM releases the inhibition of the reductase producing an increased diversion
of other folate co-factors into 5-methyltetrahydrofolate. The resultant methyl
groups re-synthesize methionine which, when activated by ATP, makes SAM.
However, this can only be accomplished by vitamin B$_{12}$-dependent methionine
synthase; thus when vitamin B$_{12}$ is deficient, 5-methyltetrahydrofolate would
continue to accumulate, producing a sort of *pseudo* folate-deficient state, i.e.
folate is trapped in a single form (5-methyltetrahydrofolate) and cannot be
used for DNA and RNA synthesis.

The methyl-trap hypothesis suggests that cells are totally unprepared for
vitamin B$_{12}$ deficiency and have not evolved to react to it effectively. When
such deficiency gives rise in turn to decreased intracellular levels of SAM, cells
respond inappropriately by increasing the synthesis of 5-methyltetrahydrofo-
late in a vain attempt to re-synthesize methionine and SAM, which in the
absence of vitamin B$_{12}$-dependent methionine synthase is impossible. Many of
the observed sequelae fit this hypothesis[8], but some do not[11]. It explains why
vitamin B$_{12}$ deficiency produces an apparently identical megaloblastic anaemia
to that seen in folate deficiency, as in both instances the cells would be deficient in the folate co-factors, 10-formyltetrahydrofolate and 5,10-methylene-
tetrahydrofolate, which are required for purine and pyrimidine biosynthesis,
respectively, and thus for DNA and RNA biosynthesis. It also explains the
clinical observation that treating a patient with megaloblastic anaemia, owing
to vitamin B$_{12}$ deficiency, with a pharmacological dose of folic acid allows the
cells of the bone marrow to start to divide again and produces a clinical haematological response. Folic acid enters the folate cycle via reduction to tetrahydrofolate, which can then be made into a polyglutamate, retained by the cell
and participate as usual in carbon one transfer for purine and pyrimidine
biosynthesis (Figure 2). Presumably, these newly added molecules of folate
would carry out several such cycles before they in turn were trapped as 5-
methyltetrahydrofolate. If folic acid continued to be administered, as it would
be if the wrong clinical diagnosis had been made, the haematological remission
would be maintained. However, since there is no endogenous synthesis of
methionine, the only way the synthesis of SAM can be maintained is by the
cell importing exogenous dietary methionine. The carbon-sulphur skeleton of
methionine used in this way would not be re-methylated back to methionine
and would be lost through catabolism (Figure 2). This could also ultimately

produce a deficiency of methionine for this purpose with a drop in the level of SAM and SAM-mediated methylation reactions. The mechanism whereby SCD, the other complication of vitamin B_{12} deficiency, is produced, appears to be both deficiency of SAM and also the toxic effects of the elevated levels of the product of SAM, namely SAH, since SAH is the main inhibitor of SAM-mediated methylation reactions. SAH levels rise, since homocysteine, the product of SAH, cannot be re-methylated when vitamin B_{12} and methionine synthase are deficient. The result is that the ratio of SAM to SAH, the so-called 'methylation ratio' is reduced, the resulting inhibition of the methylation reactions causes a state of hypo-methylation in the brain, which in turn causes SCD[12]. The neuropathy seen in SCD was thought to be exclusive to vitamin B_{12} deficiency; however, recently, a similar neuropathy has been shown to occur in patients with AIDS. While such patients are not vitamin B_{12} deficient, it has been suggested that this lesion is also owing to accumulation of SAH (possibly owing to inhibition of SAH hydrolase), causing in turn a reduced methylation ratio with inhibition of neural methyltransferase leading eventually to the neuropathy[13].

Summary

- *Folate deficiency causes anaemia owing to impaired purine and pyrimidine biosynthesis.*
- *Vitamin B_{12} deficiency causes an identical anaemia owing to metabolic trapping of intracellular folate.*
- *Vitamin B_{12} also causes nerve damage.*
- *Unlike the anaemia, the nerve damage is probably owing to inhibition of methyltransferase involved in nerve cell biosynthesis.*
- *This decreased activity is owing to a reduction in the availability of SAM, the methyl donor, together with product inhibition by its demethylated form, SAH.*

References

1. Benkovic, S.J. & Blakley, R.L. (1984) Folates and Pterins *Chemistry and Biochemistry of Folates,* vol. 1, John Wiley and Sons, New York
2. Scott, J.M. & Weir, D.G. (1976) Folate composition, synthesis and function in natural materials. *Clin. Haematol.* **5**, 547-568
3. Chanarin, I. (1979) *The Megaloblastic Anaemias,* 2nd edn., Blackwell Scientific, Oxford
4. McPartlin, J., Halligan, A., Scott, J.M., Darling M. & Weir, D.G. (1993) Accelerated folate breakdown in pregnancy *Lancet,* **341**, 148-149
5. The MRC vitamin Study Group (1991) Prevention of neural tube defects: results of the Medical Research Council vitamin study *Lancet* **338**, 131-137
6. Mathews, C.K. & Van Holde, K.E. (1990) *Biochemistry,* pp. 698-702, The Benjamin/Cummings Publishing, Redwood City, California

7.   Marcoullis, G. & Rothenberg, S.P. (1983) Macromolocules in the Assimilation and Transport of Cobalamin, in *Nutrition in Hematology* (Lindenbaum, J., ed.), chap. 3, pp. 59-119, Churchill Livingstone, New York

8.   Scott, J.M. & Weir, D.G. (1981) The methyl folate trap. A physiological response in man to prevent methyl group deficiency in kwashiorkor (methionine deficiency) and an explanation for folic acid-induced exacerbation of sub-acute combined degeneration *Lancet* **ii**, 337-340

9.   Kaufman S. (1991) Some metabolic relationships between biopterin and folate: implications for the 'methyl trap hypothesis'. *Neurochem. Res.* **16**, 1931-1936

10. Kutzbach, C. & Stokstad, E.L.R. (1971) Mammalian methylene tetrahydrofolate reductase partial purification properties and inhibition by S-adenosylmethionine. *Biochim. Biophys. Acta* **250**, 459-477

11. Charanin, I., Deacon, R., Lumb, M. & Perry, J. (1992) Cobalamin and folate: recent developments. *J. Clin. Pathol.* **45**, 277-293

12. Weir, D.G., Keating, S., Molloy, A., McPartlin, J., Kennedy, S., Blanchflower, J., Kennedy, D.C., Rice, D. & Scott, J.M. (1988) Methylation deficiency causes B_{12}-associated neuropathy in the Pig. *J. Neurochem.* **51**, 1949-1952

13. Keating, J.N., Trimble, K.C., Mulcahy, F., Scott, J.M. & Weir, D.G. (1991) Evidence of brain methyltransferase inhibition and early brain involvement in HIV-positive patients. *Lancet* **338**, 935-959

6

Protein kinase inhibitors

Hiroyoshi Hidaka and Ryoji Kobayashi

Department of Pharmacology, Nagoya University School of Medicine, Showa-ku, Nagoya 466, Japan

Introduction

Protein phosphorylation is an important cellular, regulatory event linked to the control and co-ordination of various biological activities such as metabolic pathways, gene transcription, membrane transport of ions, cell division, synaptic transmission, etc.[1-5]. The functions of cyclic nucleotides, Ca^{2+}, and diacylglycerol seem to be manifest through the phosphorylation of proteins by cyclic-nucleotide-dependent protein kinases, Ca^{2+}/calmodulin (CaM)-dependent protein kinases and by protein kinase C. While our comprehension of the biochemistry and molecular biology of protein kinases has progressed, the function of these enzymes in intact cells has been more difficult to understand. Since 1982, we have directed much attention to the biological significance of protein phosphorylation, using selective chemical inhibitors as pharmacological tools. To elucidate the physiological function of each protein kinase inhibitors should meet the following criteria: (a) direct binding to the protein kinase; (b) strict specificity for a protein kinase, and (c) cell-membrane permeability. Such inhibitors, if available would promote understanding of the role of second-messenger-related protein kinases in cellular responses [6-9].

Ca^{2+}/CaM protein kinase II

In 1980, we reported the existence of a Ca^{2+}/CaM-dependent protein kinase which phosphorylated myelin basic protein in the rabbit brain. In the same year, Yamauchi and Fujisawa partially resolved at least three CaM-dependent protein kinase in rat brain cytosol by gel filtration on Sepharose 4B, termed CaM-dependent protein kinases I, II and III. Similarly, Schulman and

Greengard found that one synaptosomal membrane protein, synapsin I, was phosphorylated by two CaM-dependent protein kinases, synapsin kinases I and II, each distinct from phosphorylase kinase and myosin light chain (MLC) kinase. Other CaM-dependent protein kinases identified during this period included an enzyme active towards glycogen synthase and phospholamban. Over the past 12 years, the above-mentioned protein kinases have been compared in detail, and it has become apparent that many of them belong to a family of closely related Ca^{2+}/CaM kinase II isoenzymes [1, 4, 10].

Ca^{2+}/CaM kinase II from rat forebrain is a large multimeric enzyme (630 kDa), composed of two related subunits with 9α (50 kDa) and 3β (60 kDa), respectively, and like other kinases undergoes an autophosphorylation that seems to be an intramolecular process. The autophosphorylation of Ca^{2+}/CaM kinase II on a threonine residue contained in a phosphopeptide common to the α- and β-subunits converts it into Ca^{2+}/CaM-independent enzyme [10]. Ca^{2+}/CaM kinase II was found to have relatively broad substrate specificities with respect to endogenous protein substrate in the nervous system. For example, phosphorylation of tyrosine hydroxylase and tryptophan mono-oxygenase by the protein kinase may alter neurotransmitter synthesis, whereas neurotransmitter release may be facilitated by phosphorylation of synapsin I. Although the enzymology of the protein kinase has been well studied, the physiological role of the enzyme remains unclear. To clarify the physiological function of Ca^{2+}/CaM kinase II, we developed specific Ca^{2+}/CaM kinase II inhibitors and eventually synthesized KN-62 [1-[N,O-bis(1,5-isoquinoline-sulphonyl)-N-methyl-L-tyrosyl]-4-phenyl-piperazine] [8, 9, 11] (Figure 1).

KN-62, a newly synthesized Ca^{2+}/CaM kinase II inhibitor[11]

Table 1 shows the effect of KN-62 on the activities of Ca^{2+}/CaM kinase II, chicken gizzard MLC kinase, cyclic AMP-dependent protein kinase and Ca^{2+}/phospholipid-dependent protein kinase (protein kinase C). More than 80% of the Ca^{2+}/CaM kinase II activity was inhibited by adding 10^{-6} M KN-62; however, the activities of the other enzymes were affected only slightly in the presence of even higher concentration of KN-62. Thus, this newly synthesized compound, KN-62, seems to be a selective and potent inhibitor of Ca^{2+}/CaM kinase II.

To elucidate the mechanisms involved in the inhibition of this kinase activity, KN-62 was tested for its ability to compete with Ca^{2+}/CaM or ATP binding to the enzyme. Figure 2 shows double-reciprocal plots of the data obtained with KN-62. Figure 2A shows the degree of inhibition obtained when the CaM concentration was varied in the absence or presence of 0.5 or 2.5 μM KN-62. As there was no change in the $V_{max.}$ value, the apparent K_m value for CaM increased with increase in the concentration of KN-62. The inhibition was competitive with respect to CaM with a K_i value of 0.9 μM. Figure 2B shows the pattern obtained when the concentration of ATP was varied. Because there was no change in the apparent K_m value when the

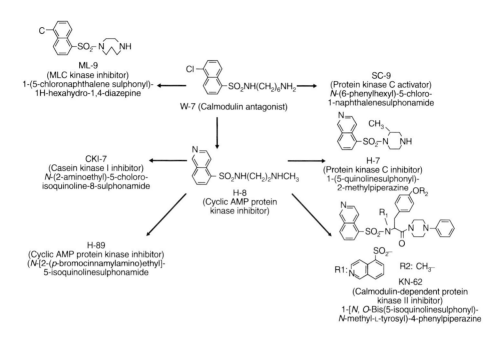

Figure 1. Chemical structures of W-7 and H-8 and the routes of development of the H-series of protein kinase inhibitors

concentrations of KN-62 were increased, KN-62 seems to inhibit kinase activity competitively with respect to CaM and non-competitively with respect to ATP.

We then examined the effect of KN-62 on autophosphorylation of the enzyme (Figure 3). Autophosphorylation of Ca^{2+}/CaM kinase II was measured in the absence of the exogenous substrate. As a result of densitometry scanning of both α- and β-subunits autophosphorylation, the sum total density of α- and β- subunits was plotted against the various concentrations of KN-62. Autophosphorylation of both α- and β- subunits of Ca^{2+}/CaM kinase II was significantly decreased by increasing the concentrations of KN-62. The IC_{50} value was between $10^{-6.5}$ and 10^{-6} M.

In the above experiments, KN-62 was added to the enzyme solution before initiation of the enzymic reaction. We then examined the effect of KN-62 on the activity of autophosphorylated Ca^{2+}/CaM kinase II. Incorporation of ^{32}P into myosin 20 kDa light chain by autophosphorylated Ca^{2+}/CaM kinase II after a 5 min reaction in the presence or absence of 50 μM KN-62 was 1.8 and 6.7 nmol of P_i/mg enzyme, respectively. We found an approximate 75% decrease in the exogenous substrate phosphorylation of Ca^{2+}/CaM kinase II when 50 μM KN-62 was added before the autophosphorylation; nevertheless, Ca^{2+}/CaM kinase II activity seen with the addition of 50 μM KN-62 after autophosphorylation at 5 min was 8.1 nmol of P_i/mg enzyme. Apparently, KN-62 does not inhibit the exogenous substrate phosphorylation

Table 1. Inhibition constants (K_I, µM) of naphthalenesulphonamides and isoquinolinesulphonamides for various kinases

[a] cyclic AMP-dependent protein kinase; [b] cyclic GMP-dependent protein kinase; [c] Ca^{2+}/CaM kinase II; [d] MLC kinase; [e] protein kinase C; [f] casein kinase I; [g] casein kinase II.

	A-kinase[a]	G-kinase[b]	CaM KII[c]	MLCK[d]	C-kinase[e]	CKI[f]	CkII[g]
Inhibitors for in vitro/in vivo use							
A-3	4.3	3.8	—	7.0	47	80	5.1
ML-9	32	—	—	4	54	—	—
H-7	3	5.8	—	97	6	100	780
H-8	1.2	0.5	—	68	15	133	950
H-88	0.4	0.8	70	50	80	60	100
H-89	0.05	0.5	30	30	30	40	140
KN-62	>100	—	0.9	>100	>100	>100	—
CKI-7	550	—	195	—	>1000	9.5	90
Affinity ligands							
H-9	19	0.9	60	70	18	110	>300
CKI-8	80	260		25	>100		

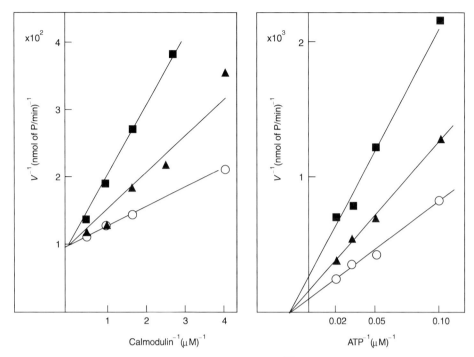

Figure 2. Double-reciprocal plots of inhibition of Ca²⁺/CaM kinase II activity by KN-62

A, rat brain Ca²⁺/CaM kinase II activity was measured in the presence of 0.24 µg/ml Ca²⁺/CaM kinase II, 1 mM CaCl₂, 10 µg of myosin 20 kDa light chain and various concentrations of calmodulin. Data are expressed as 1 (nmol/min)⁻¹ versus calmodulin 1 (µM) in the absence (○) and presence of 0.5 µM (△) and 2.5 µM KN-62 (□). B, same assay conditions as A, except calmodulin was 0.1 µM and the ATP concentration was varied in the absence (○) and presence of 0.5 µM (△) and 2.5 µM of KN-62 (□).

activity of autophosphorylated Ca²⁺/CaM kinase II. The effects of KN-62 on the CaM-independent activity of Ca²⁺/CaM kinase II were measured in the presence of the calcium chelator, EGTA. KN-62 had no effect on CaM-independent activity of the enzyme.

To demonstrate that KN-62 binds directly to Ca²⁺/CaM kinase II, sample solutions containing 73 µg of the enzyme and 12 µg of CaM in the presence of EGTA was applied to a KN-62-coupled Sepharose 4B column, and the unretarded fractions and a fraction eluted with 8 M urea were analysed by SDS/PAGE. CaM was eluted in the unretarded fractions, but Ca²⁺/CaM kinase II did not appear in these fractions. The enzyme was only eluted by boiling in SDS/PAGE sample buffer containing 8 M urea from the affinity column, suggesting that KN-62 binds directly to Ca²⁺/CaM kinase II.

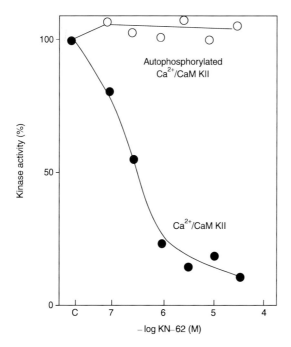

Figure 3. Effects of KN-62 on CaM-independent activity (autophosphorylated) of Ca^{2+}/CaM kinase II

Effect of KN-62 on the A23187-induced phosphorylation of Ca^{2+}/CaM kinase II in PC12 D Cells 11

To elucidate the function of Ca^{2+}/CaM kinase II in intact cells, we examined the effect of KN-62 on the activity of Ca^{2+}/CaM kinase II in PC12D cells which were immunoprecipitated with the anti-Ca^{2+}/CaM kinase II antibodies. PC12 D cells were labelled with ^{32}P-orthophosphate for 1 h. Then the calcium ionophore, A23187, was added to the cultures at a final concentration of 0.5 μM in the absence or presence of various concentrations (0.5 and 1 μM) of KN-62. After a 30 min incubation, the cell monolayers were solubilized and the cell extracts were immunoprecipitated with the anti-Ca^{2+}/CaM kinase II antibodies coupled to Sepharose 4B. Immunoprecipitated phosphoproteins were resolved by 10% (w/v) SDS/PAGE and visualized by autoradiography. Figure 4 shows that the Ca^{2+}-ionophore markedly increased the phosphorylation of the 53 kDa protein. However, the phosphorylation of the 40 kDa protein was unchanged by Ca^{2+} mobilization in the presence of KN-62. Thus, the 40 kDa phosphoprotein may be a cross-reactive protein with the antibodies, but not with Ca^{2+}/CaM kinase II. These results are consistent with the idea that the 53 kDa protein is a subunit of Ca^{2+}/CaM kinase II in PC12 D cells. Furthermore, autophosphorylation of the immunoprecipitated 53 kDa subunit of Ca^{2+}/CaM kinase II induced by A23187 was inhibited by treatment with KN-62 dose dependently. These results suggested that KN-62 was able to permeate the cell and block the Ca^{2+}/CaM kinase II activity in PC12 D cells.

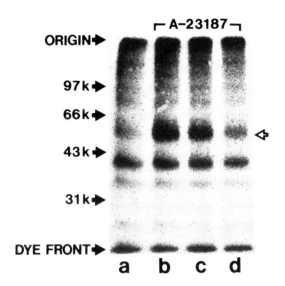

Figure 4. Effects of KN-62 on A-23187-induced phosphorylation of Ca^{2+}/CaM kinase II in PC12D cells

Tyrosine hydroxylase[12]

Activation of catecholamine-containing neurons triggers both the secretion of catecholamine and an acceleration of catecholamine biosynthesis to replenish endogenous stores that have been lost via secretion. This increase in catecholamine biosynthesis may result from the phosphorylation and activation of tyrosine hydroxylase. Tyrosine hydroxylase [TH[1]; EC 1.14.16.2; L-tyrosine, tetrahydropterine:oxygen oxidoreductase (3-hydroxylating)] is the initial enzyme in the biosynthesis of catecholamines, and its activity appears to be largely responsible for limiting the rate of this process. Recently, TH has been shown to be a substrate *in vitro* for a number of protein kinases, such as cyclic AMP-dependent protein kinase (protein kinase A), protein kinase C, Ca^{2+}/CaM kinase II and cyclic GMP-dependent protein kinase, and the phosphorylation of TH by each of these protein kinases can lead to an increase in the catalytic activity of TH. However, the question as to which kinase mainly regulates TH activity *in vivo* has been controversial.

Ca^{2+}/CaM kinase II is most highly concentrated in neural tissues, where it can represent as much as 2% of total protein in rat hippocampus. Ca^{2+}/CaM kinase II phosphorylates TH primarily at Ser-19 and slightly at Ser-40. Phosphorylation by Ca^{2+}/CaM kinase II does not affect the K_m for either pterin co-factors or tyrosine substrate. An activator protein (14-3-3 protein) activates TH 2-3-fold in $V_{max.}$ only when it has previously been phosphorylated by Ca^{2+}/CaM kinase II *in vitro*.

Although the regulation of TH *in vivo* by phosphorylation has been studied extensively, little is known about the protein kinases which phosphorylate and activate TH *in vivo*. In PC12h cells, it has been reported that Ca^{2+}/CaM kinase II, protein kinase C, and/or proline-directed protein kinase (PDPK) are responsible for the TH activation that results from depolarization in response to elevated extracellular K^+. In the rat striatum, it has been reported that Ca^{2+}/CaM kinase II or protein kinase C phosphorylate TH in response to high K^+ depolarization.

We have shown that the enhancements of both TH phosphorylation and DOPA (3-hydroxy-L-tyrosine) formation in the high K^+ depolarization were inhibited by a specific Ca^{2+}/CaM kinase II inhibitor, KN-62. On the other hand, KN-62 did not directly affect the TH activity of purified rat TH and did not reduce the phosphorylation and TH activation via protein kinase A. These results suggest that Ca^{2+}/CaM kinase II mainly phosphorylates and activates TH in the high K^+ depolarization.

Although enhancement of TH phosphorylation occurred within 30 s of incubation with 56 mM of K^+, TH activation was not observed until incubation had been continued for 3 min (Table 2). Such a time lag between phosphorylation and activation has also been reported by others. Although Ca^{2+}/CaM kinase II phosphorylates TH in response to depolarization by high K^+, some other factor(s), for example, 14-3-3 protein, an activator protein of TH, and tryptophan hydroxylase may be concerned with the subsequent activation of TH.

It has been suggested that not only Ca^{2+}/CaM kinase II, but also PDPK are responsible for the phosphorylation and activation of TH during the high K^+ depolarization. PDPK mainly phosphorylates Ser-8 and Ser-31 of rat TH. Such phosphorylation has been observed after stimulation of PC12 cells for 10 min with 60 mM K^+ but these conditions would not be physiological. Under milder conditions it appears that the phosphorylation of Ser-19 by CaM-kinase II is mainly responsible for TH activation in the high K^+ depolarization. However, a specific PDPK inhibitor would be necessary to determine whether or not PDPK also participates in the phosphorylation and activation of TH *in vivo*.

Table 2. Effect of KN-62 on TH activation by various stimuli

TH activity of PC12h cells at pH 7.0 (pmol of DOPA formed $min^{-1} \cdot mg^{-1} \cdot protein$)

	KN-62 (-)	KN-62 (10 µM)
Control	683±41	683±46
56 mM K^+ (30 s)	698±21	622±45
56 mM K^+ (3 min)	931±22	563±4.5
10 µM forskolin (10 min)	960±40	940±37

Cholinergic-stimulated parietal cell secretion[13]

The gastric parietal cell maintains a complex system for stimulation and modulation of secretion through a number of exogenous regulators. Cholinergic agonists are potent strong stimulators of acid secretion *in vivo*. Muscarinic cholinergic agonists elicit a transient stimulation of gastric secretion in both rabbit gastric glands and isolated parietal cells. In both isolated rabbit and canine parietal cells, cholinergic-activated secretion appears to be mediated by a Ca^{2+}-dependent process, and an increase in intracellular Ca^{2+} in response to the cholinergic agonist carbachol has been demonstrated with cell-permeant fluorescent dyes. The increase in intracellular Ca^{2+} is likely to be mediated by an increase in intracellular inositol trisphosphate (IP_3). However, the events mediating the stimulatory effects of elevations in intracellular Ca^{2+} levels remain obscure. Putative inhibitors of CaM, such as trifluoperazine, have been reported to inhibit parietal cell secretion. However, no studies to date have directly addressed the role or function of Ca^{2+}/CaM kinase II in parietal cells. We have therefore studied the ability of KN-62 to inhibit secretagogue-stimulated parietal cell secretion.

Rabbit parietal cell secretion is stimulated directly by both histamine and carbachol. Histamine-stimulated secretion is mediated by an increase in intracellular cyclic AMP and presumably an activation of protein kinase A. Carbachol elicits no increase in intracellular cyclic AMP, but does stimulate an increase in intracellular Ca^{2+}. However, the subsequent events following the increase of intracellular Ca^{2+} concentrations have remained obscure. It has been reported that the carbachol stimulation results in the phosphorylation of two parietal cell proteins, one of which is also phosphorylated in the presence of phorbol esters. Nevertheless, phorbol esters appear to exert complex effects on parietal cell secretion and putative protein kinase C activation does not mimic the action of carbachol. Non-specific phenothiazine CaM antagonists have been demonstrated to inhibit rat parietal cell secretion, yet the precise role of particular CaM-dependent processes remains obscure.

The results of our investigations in rabbit parietal cells indicate that an inhibitor of Ca^{2+}/CaM kinase II, KN-62, could completely inhibit carbachol-stimulated secretion. Furthermore, KN-62 failed significantly to inhibit either protein kinase A or protein kinase C *in vitro* at concentrations up to 100 μM. In similar fashion, KN-62 had no effect on cyclic AMP-mediated parietal cell secretion stimulated by either histamine or forskolin. In addition, KN-62 failed to inhibit the carbachol-stimulated increase in intracellular Ca^{2+}. All of these results are compatible with the hypothesis that Ca^{2+}/CaM kinase II is a mediator of the cholinergic stimulatory signal. These data represent, to our knowledge, the first evidence that CaM kinase II directly mediates secretion in a specific cell system.

Inhibitory effects of KN-62 on serotonin (5-HT)-evoked Cl⁻ current and ^{36}Cl⁻ efflux in *Xenopus* oocytes[14]

The *Xenopus laevis* oocyte has been used as a convenient model system in which some functional receptors are effectively reconstituted on to the plasma membranes after injection of exogenous mRNA. The effects of KN-62 on the 5-HT-induced current response and Cl⁻ efflux in *Xenopus* oocytes injected with rat brain mRNA has been investigated in this system (Figure 5).

5-HT evoked inward current responses in *Xenopus* oocytes cultivated for 20–30 h after injection of rat brain mRNA. Oocytes not injected with mRNA did not respond to 5-HT at all. The transient current in responses to 5-HT was concentration-dependent (EC_{50}: 10 nM). A second application of 5-HT 10 min after the first application induced a similar current wave response, but the peak amplitude of the transient current was decreased. This desensitization persisted for more than 1 h. Perfusion of 0.01–10 μM KN-62 for 10 min reduced the 5-HT-evoked current in a concentration-dependent manner. The application of 1 μM KN-62 also inhibited the currents resulting from the application of 1 mM acetylcholine (ACh) in mRNA-injected oocyte and also the current evoked by intracellular injection of 50 pmol of IP_3 in the native oocyte.

It has been suggested that the 5-HT-evoked inward current in the oocyte is due to Cl⁻ efflux, based on the reversal potential and the inhibition of this

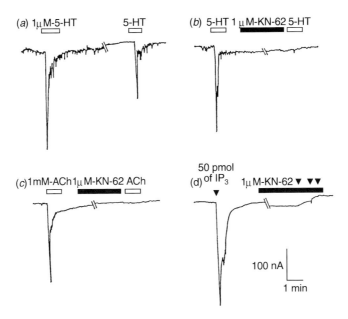

Figure 5. Effects of KN-62 on 5-HT (b), ACh-(c) and IP_3 (d)-induced current response
Transmembrane currents were measured by voltage-clamp methods, holding at -60 mV. Each oocyte was given 5-HT or ACh twice. The first: baseline control; the second: following 10 min perfusion with KN-62 (b-d) or vehicle (a).

current by a Cl⁻-channel inhibitor, 4,4′-di-isothiocyanostilbene-2,2′-disulphonic acid. Thus the effects of KN-62 on 5-HT-induced ^{36}Cl⁻ efflux from the oocyte injected with mRNA were also examined. 5-HT significantly enhanced ^{36}Cl⁻ efflux from the oocyte during 1 min, and KN-62 reduced the 5-HT response in a concentration-dependent manner.

KN-62 potently inhibited the 5-HT-, ACh- and IP$_3$-induced Cl⁻ current, and also inhibited the 5-HT-induced Cl⁻ efflux. These results suggest that Ca^{2+}/CaM kinase II is involved in the signal transduction mechanisms from receptor stimulation to Cl⁻ efflux in *Xenopus* oocytes.

Effects of KN-62 on long-term potentiation in the rat hippocampus[15]

Long-term potentiation (LTP) of synaptic transmission in the hippocampus is considered as a cellular model of learning and memory. In this process rapidly repeated (tetanic) stimulation of glutamatergic nerves results in a long-term increase in the sensitivity of the system to subsequent stimuli. Extensive studies have revealed the presence of at least two forms of LTP with different cellular mechanisms. One is LTP in the CA1 region of the brain. The induction of this LTP is dependent upon the *N*-methyl-D-aspartate (NMDA) class of glutamate receptors. The other is LTP of mossy fibre-CA3 pyramidal cell synapses, in which the involvement of NMDA receptors is much less important.

Studies on the LTP in CA1 regions have suggested that activation of both protein kinase C and Ca^{2+}/CaM kinase II is required for the induction of LTP. However, the involvement of protein kinase C in the maintenance of LTP has recently been questioned.

We have used KN-62, and analysed the physiological roles of Ca^{2+}/CaM kinase II in the hippocampal LTP. This compound is effective when applied in extracellular media at micromolar concentrations, and inhibits Ca^{2+}/CaM kinase II specifically without affecting other kinases.

Experiments were performed using rat hippocampal slices. Figure 6 shows the effects of KN-62 on Schaffer/commissural-CA1 pyramidal synapses. The compound has no significant effects on the transmission itself, but strongly suppresses the LTP when applied before the tetanic stimulation (Figure 6a). The relative amplitude of population excitatory post-synaptic potential (pEPSP) measured 30 min after the tetanus was 144±3% (n = 6) of the value immediately before the tetanus for the control slices, whereas it was 104±4% (n = 7) for slices pretreated for 20 min with 3 μM KN-62 (mean±S.E.M.).

When KN-62 was applied in the perfusate immediately after the tetanic stimulation, however, the LTP was observed essentially to the same extent as control (Figure 6b). The relative amplitude of pEPSP was 146±9% (n = 5). These results indicate that the activation of Ca^{2+}/CaM kinase II is required for LTP generation, and further show that the activation is critical only in the initial phase of LTP generation in CA1 regions.

In contrast to the LTP in CA1 regions, however, the LTP of mossy fibre-

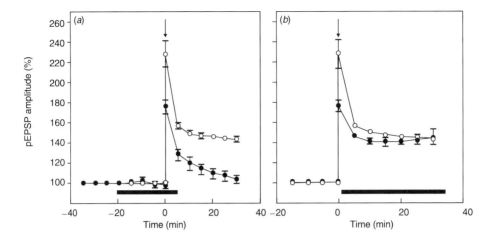

Figure 6. Effects of KN-62 on LTP in the CA1 region
KN-62 (3 μM) was applied (thick bars) (a) before or (b) immediately after the tetanic stimulation
(arrows). The data indicate means±S.E.M. The numbers of experiments were (a) six for control
(○) and seven for KN-62 (●), or (b) six for control (○) and five for KN-62 (●).

CA3 pyramidal cell synapses is much less sensitive to KN-62. The relative
amplitudes of pEPSP measured 30 min after the tetanic stimulation were
154±5% (n = 5) and 148±8% (n = 4) for the control and experimental slices,
respectively.

The difference in the sensitivity to KN-62 between the LTP in CA1
regions and the LTP of mossy fibre synapses in CA3 regions apparently
resembles the difference in the sensitivity to 2-amino-5-phosphonovalerate
(AP5), an NMDA antagonist, between these two types of LTP. Thus it is
possible that KN-62 might interfere with some NMDA receptor functions
similarly to AP5. To make this point clear, we examined effects of KN-62 on
NMDA receptors expressed in *Xenopus* oocytes injected with rat brain
mRNA. Maximum amplitudes of current responses evoked by 100 μM
NMDA were measured under voltage-clamp conditions in external media
containing 3 μM glycine and no Mg^{2+}. The relative amplitude in the presence
of 3 μM KN-62 was 92±4% (mean±S.E.M.; n = 5) of the value in its absence,
indicating that KN-62 has no significant effects on NMDA receptor functions.
Thus the inhibitory effects of KN-62 on CA1 LTP probably result from its
direct action on Ca^{2+}/CaM kinase II, and not to non-specific side-effects.

The significance of protein kinases in mossy fibre LTP has been very
poorly understood, and the present results suggested that Ca^{2+}/CaM kinase II
may be much less important in mossy fibre LTP than in CA1 LTP.

Long-term depression is induced in Ca^{2+}/CaM kinase II-inhibited visual cortex neurons[16]

In the visual cortex in particular, long-term depression (LTD) as well as LTP is supposed to be a basis for environmental modifiability of cortical neurons during a 'critical period' of postnatal development. Recently, it has been suggested that the level of post-synaptic depolarization or post-synaptic Ca^{2+}-concentration may be critical for determining the direction of the long-term changes following tetanic inputs. After inhibition of post-synaptic Ca^{2+}/CaM kinase II by direct injection of its inhibitor, KN-62, into layer 2/3 neurons in sliced visual cortex, tetanic stimulation of the white matter induced LTD of the excitatory post-synaptic potentials (EPSPs) at tetanized synapses, while the EPSPs at non-tetanized synapses remained unchanged. The majority of control cells which were recorded with KN-62-free electrodes showed LTP of the EPSPs. This suggests that inhibition of post-synaptic Ca^{2+}/CaM kinase II during tetanic synaptic input leads to LTD, while its activation leads to LTP.

Effect of KN-62 on MLC kinase phosphorylation[17]

Phosphorylation of the regulatory light chain of myosin by Ca^{2+}/CaM-dependent MLC kinase plays a primary role in initiating smooth muscle contraction. Although cytosolic Ca^{2+} concentrations are a primary determinant for MLC phosphorylation, it has become apparent that other factors may regulate the Ca^{2+} dependence of MLC phosphorylation in smooth muscle cells.

Recent studies have demonstrated that MLC kinase is phosphorylated to a high extent at a regulatory site (site A) in bovine tracheal smooth muscle when the muscle is stimulated to contract with carbachol or KCl[18].

The transient increase in the extent of Ca^{2+}-dependent MLC phosphorylation observed with carbachol stimulation of tracheal smooth muscle tissues was paralleled temporally with a transient increase in site A phosphorylation. Activity ratios varied linearly as a function of the cytosolic free Ca^{2+} concentration ($[Ca^{2+}]_i$) in strips stimulated with KCl or carbachol. In Ca^{2+}-depleted tissues, MLC and MLC kinase phosphorylation were inhibited. These findings led us to propose that phosphorylation of MLC kinase at regulatory site A occurs via a Ca^{2+}-dependent kinase in intact smooth muscle. Purified MLC kinase is phosphorylated at site A by two Ca^{2+}-dependent enzymes, protein kinase C and Ca^{2+}/CaM kinase II. It is unlikely that protein kinase C catalyses the phosphorylation of this site in intact muscle, however, since treatment of tracheal strips with a phorbol ester does not result in significant phosphorylation of regulatory site A. Thus, Ca^{2+}/CaM kinase II appears to be the more likely candidate.

Stimulation of tracheal smooth muscle cells in culture with ionomycin resulted in a rapid increase in $[Ca^{2+}]_i$ and an increase in both MLC kinase and MLC phosphorylation. These responses were markedly inhibited in the absence of extracellular Ca^{2+}. Pretreatment of cells with KN-62 did not affect the increase in $[Ca^{2+}]_i$, but inhibited ionomycin-induced phosphorylation of MLC kinase at the regulatory site near the calmodulin-binding domain. KN-62 inhibited Ca^{2+}/CaM kinase II activity towards purified MLC kinase.

Phosphorylation of MLC kinase decreased its sensitivity to activation by Ca^{2+} in cell lysates. Pretreatment of cells with KN-62 prevented this densitization to Ca^{2+} and potentiated MLC phosphorylation. We propose that the Ca^{2+}-dependent phosphorylation of MLC kinase by Ca^{2+}/CaM kinase II decreases the Ca^{2+} sensitivity of MLC phosphorylation in smooth muscle.

Ca^{2+}/CaM kinase II and protein kinase C are required for endothelin-stimulated atrial natriuretic factor secretion[19]

Endothelin (ET)-stimulated atrial natriuretic factor (ANF) secretion from atrial myocytes can be divided into two components which differ in their requirements for Ca^{2+} influx. The Ca^{2+}-influx-dependent component, which is defined as that portion of ET-stimulated ANF secretion that can be inhibited with the calcium channel blocker nifedipine in Ca^{2+}-free medium, accounts for about 50% of ET-stimulated ANF secretion. The Ca^{2+}-influx-independent component is defined as the portion of ET-stimulatable ANF secretion that remains after treatment with nifedipine or Ca^{2+}-free medium. We have investigated the intracellular mechanisms underlying these two components of ET-stimulated ANF release and demonstrated that protein kinase C and Ca^{2+}/CaM kinase II mediate the Ca^{2+}-influx-independent and -dependent portions of secretion, respectively. The protein kinase C inhibitor, H7, blocked only Ca^{2+}-influx-insensitive secretion, while the combination of H7 and nifedipine completely blocked stimulated secretion. Secondly, ET-mediated phosphorylation of the protein kinase C substrate protein p80 was unaffected by blocking Ca^{2+} influx with nifedipine; in contrast to ET, the compound BAY K 8644, which also stimulates ANF release, did not stimulate p80 phosphorylation.

   Our findings further indicated that the Ca^{2+}-influx-sensitive component of ET-stimulated ANF secretion involved Ca^{2+}/CaM kinase II. First, KN-62 blocked the Ca^{2+}-influx-sensitive component, like nifedipine, and its effects were additive with H7 such that both compounds completely abolished ET-induced secretion. Secondly, BAY K 8644-stimulated ANF release, which has been shown to be inhibited by nifedipine, was sensitive to KN-62 (ANF secretion with 1 µM BAY K = 4.40±0.96-fold over control; 1 µM BAY K +5 µM KN-62 = 1.50±0.18-fold over control). These results indicated that maximal ET-mediated ANF secretion involved the activities of both protein kinase C and Ca^{2+}/CaM kinase II.

KN-93, a water-soluble Ca^{2+}/CaM kinase II inhibitor[20]

KN-62 is a potent and specific inhibitor for CaM kinase II and is cell-permeable and hydrophobic, but these properties make for difficult application in *in vivo* or *in situ* experiments. To use KN-62 in such systems it must be dissolved in dimethylsulphoxide (DMSO) and this sometimes harms some tissues or cultured cells. Because of this, we attempted to develop a water-soluble Ca^{2+}/CaM kinase II inhibitor. We have recently investigated the inhibitory property of a newly synthesized methoxybenzenesulphonamide, KN-93, on Ca^{2+}/CaM kinase II activity *in situ* and *in vitro*. KN-93 elicited potent inhibitory effects on Ca^{2+}/CaM kinase II phosphorylating activity with an inhibition constant of 0.37 μM, but this compound had no significant effects on the catalytic activity of protein kinase A, protein kinase C, MLC kinase and Ca^{2+}-phosphodiesterase. KN-93 also inhibited the autophosphorylation of both the α- and β-subunits of Ca^{2+}/CaM kinase II. Kinetic analysis indicated that KN-93 inhibits Ca^{2+}/CaM kinase II, in a competitive fashion against CaM. To evaluate the regulatory role of Ca^{2+}/CaM kinase II on catecholamine metabolism, we examined the effect of KN-93 on dopamine (DA) levels in PC12h cells. The DA levels decreased in the presence of KN-93. Further, the TH phosphorylation induced by KCl or acetylcholine was significantly suppressed by KN-93 in PC12h cells, while events induced by forskolin or 8-Br-cyclic AMP were not affected. These results suggest that KN-93 inhibits DA formation by modulating the reaction rate of Ca^{2+}/CaM kinase II to reduce the Ca^{2+}-mediated phosphorylation levels of the TH molecule.

The most noteworthy advantage of KN-93 is its water-solubility. Solvents used for water-insoluble chemicals, such as DMSO or ethanol, cannot always be used for physiological assays. When the free-form of KN-93 in DMSO as the vehicle was used in similiar experiments to those described above the potency was exactly the same as that of the water-soluble form. Thus KN-93 should become a useful and convenient pharmacological tool for elucidating physiological functions of Ca^{2+}/CaM-kinase II in intact cells.

MLC kinase inhibitor

Diverse Ca^{2+}-dependent cellular processes, such as smooth muscle contraction and non-muscle cell secretion, are thought to be regulated by Ca^{2+}-dependent protein phosphorylation systems.

Ca^{2+}-dependent MLC phosphorylation is catalysed by CaM-dependent MLC kinase and protein kinase C. MLC kinase is apparently phosphorylated by protein kinase C and by protein kinase A[18].

As CaM activates many enzymes, Ca^{2+} dependently, it is difficult to evaluate which CaM-dependent enzyme is mainly responsible for the Ca^{2+}-dependent regulatory system in smooth muscle and non-muscle cells, using the CaM antagonist, W-7[6-9].

On the other hand, ML-9 has selective inhibitory effects on MLC kinase of smooth muscle and platelets, but is less potent in inhibiting skeletal muscle MLC kinase, other CaM-dependent enzymes, protein kinase C and protein kinase A. These results suggest that ML-9 is a specific inhibitor of MLC kinase and hence will be a useful tool for investigating the function of MLC kinase in smooth muscle and non-muscle cells (Figure 1).

When the mode of interaction of ML-9 with MLC kinase was examined by kinetic analysis, we found that it was a competitive inhibitor versus ATP and a non-competitive inhibitor versus the phosphate acceptor, in the catalytic reaction. ML-9 is not structurally related to ATP nor could it serve as a substrate for the kinase. Therefore, it is likely that ML-9 blocks the enzyme activity by interacting with the free enzyme, but not with the enzyme-ATP complex. These results indicate that ML-9 binds at or near the ATP-binding site at the active centre of the kinase resulting in inhibition of the catalytic activity of MLC kinase.

Smooth muscle contraction

The addition of 20 mM KCl caused a sustained contraction in rabbit aortic strips. ML-9 produced a dose-dependent relaxation, in a 5.8 µM concentration which produced a 50% relaxation of sustained contraction ($n = 5$). This is similar to the K_i value of ML-9 in kinase. We also investigated the effect of ML-9 on the Ca^{2+}-induced contraction in chemically skinned muscles from rabbit mesenteric arteries to exclude the possibility of inhibition of the transmembrane Ca^{2+} influx. ML-9 significantly suppressed the Ca^{2+}-induced contraction of the skinned fibres and MLC phosphorylation.

Platelet secretion

To elucidate the physiological function of MLC phosphorylation in human platelet secretion, the effects of ML-9 on the release reaction and protein phosphorylation of washed human platelets preincubated with ^{32}P-orthophosphate of ^{14}C-5-HT and treated with collagen (1 µg/ml) were investigated.

When human platelets were activated by collagen, myosin 20 kDa light chain phosphorylation appeared rapidly to precede shape change. Secretion, aggregation and phosphorylation of the 40 kDa peptide followed. Two-dimensional peptide mapping of this phosphorylated MLC induced by collagen, following tryptic digestion, revealed that only MLC kinase phosphorylated the light chain, and the site phosphorylated by protein kinase C was not detected in the early phase during collagen activation. ML-9 delayed the time course of the MLC phosphorylation, in a dose-dependent manner. We also observed delays in 5-HT secretion, platelet aggregation and phosphorylation of the 40 kDa peptide. ML-9 had no effect on protein phosphorylation and secretion in unstimulated platelets. We also investigated the effect of ML-9 on the 40 kDa peptide phosphorylation induced by other agonists, to clarify that the delay in the 40 kDa peptide phosphorylation is not due to the direct effect of ML-9 on protein kinase C. ML-9 did not suppress the phosphorylation of either the 40 or 20 kDa peptides induced by12-O-tetradecanoylphorbol 13-acetate (TPA), in the same concentration range of ML-9.

The effect of ML-9 was also examined on Ca^{2+}-dependent phosphorylation of the MLC and Ca^{2+}, phospholipid-dependent phosphorylation of the 40 kDa peptide, in a crude cell-free system. Increasing the concentration of ML-9 inhibited selectively the Ca^{2+}-dependent MLC phosphorylation, with an IC_{50} value of 12 µM. However, ML-9 had little effect on Ca^{2+}, phospholipid-dependent phosphorylation of the 40 kDa peptide, in concentrations up to 50 µM, at the concentration of which, this compound completely inhibits human platelet secretion. These observations suggest that MLC phosphorylation catalysed by MLC kinase may have a critical role in a series of reactions, such as secretion, aggregation and phosphorylation of the 40 kDa peptide, in response to the stimulation of human platelets by collagen.

Protein kinase C[7-9]

Protein kinase C is a family of closely related monomeric protein kinases. Since the initial identification of protein kinase C, pharmacological approaches have been used to elucidate the physiological roles of the enzyme. Related activators include proteases, Ca^{2+}, phospholipids, diacylglycerols, phorbol esters, and SC-9 and SC-10[21]. These compounds, in particular the phorbol esters, have frequently been used as tools to elucidate the physiological role of protein kinase C. However, it is difficult to prove the physiological significance of the enzyme without also suppressing the enzyme activity using inhibitors. CaM antagonists, polypeptides, polyamines, local anaesthetics, doxorubicin (adriamycin) and other lipids are lipophilic and interact with phospholipids, resulting in an indirect inhibition of protein kinase C. Moreover, these inhibitors are relatively non-selective in their action on phospholipid/Ca^{2+}- and CaM/Ca^{2+}-dependent enzymes. Since protein kinase C is present in almost all eukaryotic cells and plays a role in many physiological and pathological processes, there is strong interest in developing specific inhibitors for protein kinase C.

At higher concentrations, the CaM antagonist, W-7, inhibits protein kinase C activity. During the synthesis and selection of CaM antagonists derived from naphthalenesulphonamide, we discovered that a shorter alkyl chain derivative, termed A-3 (N-[6-aminoethyl]-5-choloro-naphthalene sulphonamide), markedly inhibited protein kinases through a mechanism different from that of W-7. When the amino residue at the end of the hydrocarbon chain of W-7 was changed to a phenyl residue, the compound (SC-9) stimulated protein kinase C activity. Replacing the naphthalene ring of naphthalene sulphonamides with isoquinoline resulted in derivatives which no longer interacted with CaM or phospholipid, but rather compounds which potently suppress protein kinase C and cyclic nucleotide-dependent protein kinases. Among the isoquinolinesulphonamides, 1-(5-isoquinolinesulphonyl)-2-methylpiperazine (H-7) exhibits a relatively selective inhibition of protein kinase C (Figure 1)[22]. Kinetic analysis revealed that the inhibition of protein kinase C by H-7 was competitive with respect to ATP and non-competitive with respect to the phosphate acceptor. The K_i value of this compound against protein kinase C is 6.0 µM. Radiolabelled isoquinolinesulphonamide derivatives are incorporated into the cells; and evidence for their binding to the ATP-binding site of a protein kinase with a binding ratio of 1:1 was obtained using the gel-permeation binding assay.

H-7 is now widely used for studies on various biological systems [7-9]. It has significant effects on various functions by inhibiting protein kinase C-induced phosphorylation in cells (Table 3). For example, in platelets, H-7 enhances 5-HT release induced by the phorbol ester TPA or thrombin, in association with inhibition of the protein kinase C-catalysed phosphorylation of the MLC. It also attenuates TPA-induced inhibition of phosphoinositide metabolism and Ca^{2+} mobilization in thrombin-activated human platelets. It prevents TPA-induced inhibition of the NaF-mediated rise in $[Ca^{2+}]$, and thromboxane B_2 generation in platelets.

The effects of TPA or H-7 on various functions of leucocytes have also been extensively investigated. Preincubation of mast cells in the presence of TPA results in an inhibition of histamine secretion induced by compound 48/80 (which appears to act primarily by raising intracellular free Ca^{2+} concentrations) to 65% of the maximal response observed in the absence of TPA pre-treatment; however, treatment of cells with H-7 prevented these changes in the secretory activity. In addition, H-7 inhibited the phosphorylation of the 36 kDa protein enhanced by protein kinase C activators, such as phosphatidylserine, phosphatidylinositol and phosphatidylethanolamine.

Use of H-7 and H-9, has demonstrated that CaM-dependent enzymes rather than protein kinase C may be responsible for neutrophil activation. H-7 inhibits the TPA-induced HTLV-1 p19 antigen expression in the HTLV-1 virus infected human T-cell line, KH-2Lo cells, and inhibits the TPA-induced reduction of vincristine uptake from p388 murine leukaemic cells, thereby suggesting its possible role as a chemotherapeutic agent. In pancreatic islet cells, H-7 causes a partial decrease in insulin release evoked by TPA. All these data clearly show that when examining the function of protein kinase C *in vivo*, the biphasic effects of both TPA and H-7 should be examined. Among the derivatives, H-9 [*N*-(2-aminoethyl)-5-isoquinolinesulphonamide] possesses a primary amino group and can be used as a ligand for affinity chromatography. Large-scale purification to homogeneity was achieved for protein kinase C by the use of the H-9 affinity column.

Cyclic-nucleotide-dependent protein kinase inhibitors, H-8, H-88, H-89[8, 9, 23]

Protein kinase A is characterized by having a broad tissue distribution as well as an involvement in diverse responses. We found that derivatives of W-7 exhibit a relatively selective inhibition toward protein kinase A and cyclic

Table 3. Effects of protein kinase C inhibitor

Inhibitor	Action	Cell type
H-7, W-7	Superoxide production	Human histocytic leukaemia cell (U937)
H-7	Antigen and interleukin (IL)-2-induced proliferation	Murine T-cell
H-7	Oxidative burst and degranulation	Human neutrophil
H-7	Glucocorticoid-dependent enzyme induction and glucocorticoid receptor translocation	Rat hepatocyte
H-7	Chemotaxis	Human neutrophil
H-7	Superoxide production and secretion	Rabbit peritoneal neutrophil
H-7	β_2-Adrenergic receptor activation and desensitization	Human lymphocyte
H-7	Cell locomotion	Human polymorphonuclear neutrophil
H-7, W-7	Atrial natriuretic peptide release	Rat atrium
H-7	Hormonal receptor-adenylate cyclase complex modulation	Human lymphocyte
H-7, polymyxin B, staurosporine	Platelet aggregation	Human platelet
H-7, HA1004	Calcitriol-induced differentiation	HL-60 cell
H-7, W-7, HA1004	Morphological change	Rat astrocyte
H-7	Neurite outgrowth	Mouse neuroblastoma and cerebellar cells
H-7	Long-term potentiation	Rat Ca1 pyramidal cell
H-7	Cell polarity and locomotion	Walker-256 cell
H-7	Alteration of actin cytoskeleton	Swiss 3T3 cell
H-7	Intracelluar actin level	HL-60 cell
H-7, W-7	Pertussis toxin-induced IL-1 production	Human monocyte
H-7	Cell lysis and proliferation	Human lymphocyte
H-7	Natural-killer activity	Human granular lymphocyte
H-7	Tumour promotion	Epidermal JB cell
H-7	Stress, cross-bridge phosphorylation, muscle shortening and inositol phosphate production	Rabbit artery
H-7, Sphingosine	Cellular proliferation	Swiss 3T3 cell

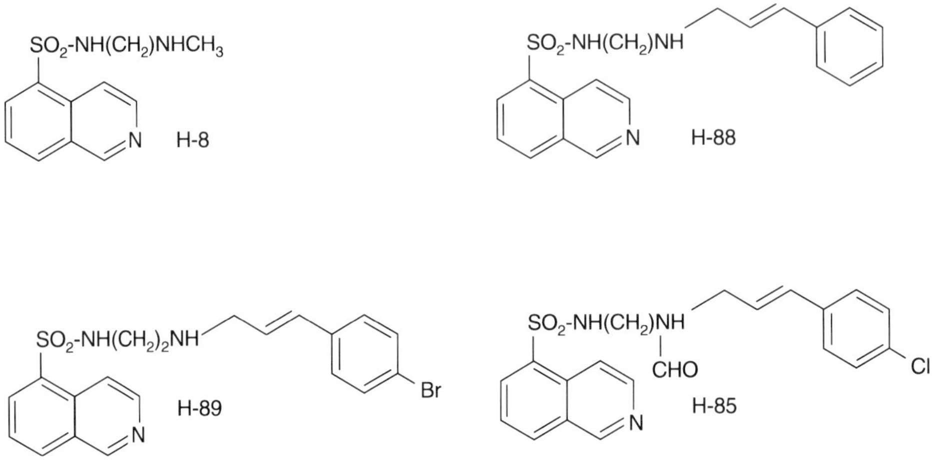

Figure 7. Structures of inhibitors of cyclic nucleotide-dependent protein kinases

GMP-dependent protein kinase, and that H-8, N-[2-(methylamino)ethyl]-5-isoquinolinesulphonamide, was a potent inhibitor of cyclic-nucleotide-dependent protein kinase (Table 1; Figure 7). The inhibition was of the competitive type with respect to ATP. We found that H-8 specifically binds to the ATP-binding site of the catalytic subunit with a binding ratio of 1:1.

A more specific inhibitor of protein kinase A is now available. A newly synthesized isoquinolinesulphonamide, H-89 (N-[2-(p-bromo-cinnamyl-amino)ethyl]-5-isoquinolinesulphonamide), which is a much more potent inhibitor of protein kinase A than H-8 (see Figure 7).

The effects of these newly synthesized isoquinolinesulphonamides on protein kinase A, cyclic GMP-dependent protein kinase, protein kinase C, MLC kinase, Ca^{2+}/CaM kinase II, and casein kinase I and II have been investigated. H-88, N-(2-cinnamylaminoethyl)-5-isoquinoline sulphonamide, had a potent inhibitory effect on both protein kinase A and cyclic GMP-dependent protein kinase (Table 1). H-89, a brominated derivative of H-88, was the most selective and potent inhibitor of cyclic AMP-dependent protein kinase among the isoquinolinesulphonamide derivatives tested. Its K_i value for protein kinase A was 0.05 µM while the K_i value for the cyclic GMP-dependent kinase was 10 times higher, suggesting that the inhibitor was specific for protein kinase A (Table 1). H-89 was shown to be a much weaker inhibitor of other kinases such as protein kinase C, MLC kinase, Ca^{2+}/CaM kinase II, and casein kinase I and II. To elucidate the mechanisms involved in the inhibition of protein kinase A activity, both H-88 and H-89 were tested for their ability to compete with ATP binding to this enzyme.

Effects of H-89 on forskolin- and nerve-growth-factor-induced phosphorylation in PC12D cells[23]

To investigate whether H-89 could serve as a pharmacological probe for examining intact cells, we studied the effect of this inhibitor on forskolin- and nerve growth factor (NGF)-induced phosphorylation in PC12D cells. In the absence of forskolin, ^{32}P-radioactive phosphorylated proteins were not detected in PC12D cells. Pre-treatment of the cells with H-89, 1 h before the addition of forskolin, markedly inhibited forskolin-induced protein phosphorylation in a dose-dependent fashion. These inhibitory effects were observed even 8 h after the addition of forskolin. The inhibition of NGF-induced protein phosphorylation was not observed in the PC12D cells pre-treated with H-89. These results suggest that H-89 is a useful probe with respect to the selective inhibition of protein kinase A in intact cells.

The effects of H-89 on forskolin-induced cyclic AMP accumulation in PC12D cells were also examined. The addition of H-89 along with forskolin did not affect forskolin-induced increase in the cyclic AMP levels in PC12D cells. H-89 did not inhibit adenylate cyclase and cyclic nucleotide phosphodiesterases, at concentration up to 100 μM. Thus, H-89 may act directly on protein kinase A in the intact PC12D cells.

Chromogranin B/secretogranin I and secretogranin II expression by forskolin[24]

The granins [chromogranin A (CgA), chromogranin B/secretogranin I (CgB), and secretogranin II (SgII)] are a family of immunologically distinct proteins stored in the matrix of secretory granules in most endocrine and neuronal cells. The physiological function of these proteins is unknown. As they are secreted, it has been suggested that they may play an extracellular role as hormone precursors.

Despite numerous studies on the structure, sequence and intracellular localization of the granins, there is little information on the regulation of the expression of these proteins. Recent studies have focused on the differential regulation of the granin family of proteins by intracellular mediators such as cyclic AMP. For example, forskolin increased steady-state levels of SgII mRNA in bovine chromaffin cells, but had no effect on the level of CgA mRNA in the same system, in the MTC medullary thyroid carcinoma cell line nor in BEN lung tumour cells.

A number of cyclic AMP-independent actions of forskolin have been described, including a direct effect of the diterpene in PC12 cells. To investigate whether the action of forskolin on CgB mRNA levels was mediated by cyclic AMP, we examined: (i) the effect of the cyclic AMP analogue, 8-Br-cyclic AMP, on CgB mRNA levels; (ii) the effect of forskolin in the presence of H-89. We found that the administration of 500 μM 8-Br-cyclic AMP increased CgB mRNA levels to 280±30% of control levels, suggesting that CgB mRNA levels are regulated by cyclic AMP in PC12D cells. In another experiment, a higher concentration of 8-Br-cyclic AMP (1 mM) increased CgB mRNA levels to the same extent as 10 µM forskolin. Pre-incubation of the cells with H-89 reduced the increase in CgB mRNA levels induced by forskolin from 550±30% to 180±50% of control levels, suggesting that forskolin induction of CgB mRNA expression is dependent on an increase in cellular cyclic AMP and activation of protein kinase A.

Casein kinase I inhibitors, CKI-6, CKI-7, CKI-8[8, 9, 25]

Casein kinase I has been purified from many tissues, including calf thymus, rabbit reticulocytes, liver and skeletal muscle. The widespread distribution of the enzyme suggests it is important in cellular function, although exactly how remains to be clarified. H-9 has a weak inhibitory effect on casein kinase I and II. When the 5-aminoethyl-sulphonamide chain of H-9 was moved to position 8 on the aromatic ring, the derivative, CKI-6 [N-(2-aminoethyl)-isoquinoline 8-sulphonamide], produced a more potent inhibition of casein kinase activity than did H-9.

CKI-7 [N-(2-aminoethyl)-5-chloroisoquinoline 8-sulphonamide, the chlorinated derivative of CKI-6] (see Figure 8), potently inhibited casein kinase I with an IC_{50} value of 9.5 µM for casein kinase I and 90 µM for casein kinase II. CKI-7 at a concentration of up to 100 µM only weakly inhibited protein kinase C. IC_{50} values of CKI-7 were 550 µM for protein kinase A and 195 µM for Ca^{2+}/CaM kinase II activity. Kinetic analyses showed that CKI-7 inhibited casein kinase I competitively with respect to ATP. CKI compounds were also used as affinity ligands to purify casein kinase I. Two kinds of affinity columns were used incorporating CKI-7 or CKI-8 [1-(5-chloro-8-isoquinolinesulphonyl)-piperazine], as affinity ligands. To compare the properties of the two affinity ligands, partially purified casein kinase I and II were applied to these affinity columns. The CKI-7 affinity column absorbed casein kinase I only and casein kinase II did not bind to it. The CKI-8 affinity column, by contrast, absorbed both casein kinase I and II. These results suggest that CKI-7 has a specific affinity for casein kinase I. CKI-8 does not have a specific affinity for casein kinase I and demonstrates an inhibition for casein kinase I that is approximately 10 times weaker than CKI-7. However, the recovery of casein kinase I from the CKI-7 column was lower than the CKI-8 affinity column, suggesting that casein kinase I was bound tightly to the column. The CKI-8 affinity column was thus chosen as the final step in purifying casein kinase I. Casein kinase I eluted from the affinity column at an L-arginine concentration between 0.53 and 0.8 M. Eighteenfold purification was achieved by this step alone. Although several casein kinase II inhibitors have been used to investigate the physiological role of the enzyme, a selective casein kinase I inhibitor has not been available. CKI-7 and CKI-8 should thus prove useful in determining the physiological role and distribution of casein kinase I in different tissues.

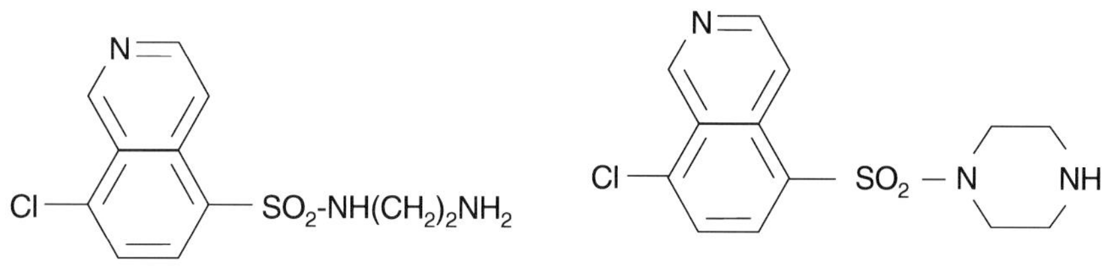

Figure 8. Structures of inhibitors of casein kinase I and II

Conclusion

A brief description was made of biochemical and pharmacological approaches which can be used to elucidate the role of multiple protein kinases in regulating various cellular functions. The number of protein kinases and their complex interactions on the same systems has made it difficult to define their separate roles in a variety of cellular processes. Selective inhibitors, the so-called H-series inhibitors, which inhibit cyclic-nucleotide-dependent protein kinases, MLC kinase, protein kinase C, Ca^{2+}/CaM kinase II and casein kinase I have now been developed by the authors for the first time. While biochemical studies on the functions of protein kinases with isolated systems *in vitro* are feasible, studies *in vivo* are relatively difficult to carry out. Thus, to clarify the physiological role and molecular mechanisms of the individual protein phosphorylation systems, selective inhibitors are proving useful as pharmacological probes in intact systems.

Related Reading

Long-term potentiation (section II-6) is discussed in detail by Rose, S.P.R. (1991) The biochemistry of memory. *Essays Biochem.* **26**, 1-12

The actions of the endothelins (section II-9) are described by Masaki, T. & Yanagisawa, M. (1992) Endothelins *Essays Biochem.* **27**, 79-89

The role of protein phosphorylation in synaptic transmission is discussed by Rodnight, R. & Wolfchuk, S.T. (1992) Roles for protein phosphorylation synaptic transmission *Essays Biochem.* **27**, 91-102

References

1.   Schulman, H. (1988) The multifunctional Ca^{2+}/calmodulin-dependent protein kinase. *Adv. Second Messengers Phosphoprotein Res.* **22**, 39-112

2.   Kikkawa, U., Kishimoto, A. & Nishizuka, Y. (1989) The protein kinase C family: Heterogeneity and its implications. *Annu. Rev. Biochem.* **58**, 31-44

3.   Taylor, S.S., Buechler, J.A. & Yonemoto, W. (1990) Cyclic AMP-dependent protein kinase: framework for a diverse family of regulatory enzymes. *Annu. Rev. Biochem.* 59, 971-1005

4.   Nairn, A.C., Hemmings, H.C. Jr & Greengard, P. (1985) Protein kinases in the brain. *Annu. Rev. Biochem.* **54,** 931-976

5.   Edelman, A.M., Blumenthal, D.A. & Krebs. E.G. (1987) Protein serine/threonine kinases. *Annu. Rev. Biochem.* **56,** 567-613

6.   Hidaka, H., Tanaka, T., Saitoh, M. & Matsushima, S. (1987) Molecular pharmacology of myosin light chain phosphorylation of smooth muscle and nonmuscle cells, in *Calcium-Binding Proteins in Health and Disease* (Norman, A.W., Vanaman, T.C. & Means, A.R., eds.) pp. 170-179, Academic Press, Orlando

7.   Hidaka, H. & Tanaka, T. (1987) Transmembrane Ca^{2+} signalling and a new class of inhibitors. *Methods Enzymol.* **139,** 570-582

8.   Hidaka, H., Watanabe, M. & Kobayashi, R. (1991) Properties and use of H-series compounds as protein kinase inhibitors. *Methods Enzymol.* **201,** 328-339

9.   Hidaka, H. & Kobayashi, R. (1992) Pharmacology of protein kinase inhibitors. *Annu. Rev. Pharmacol. Toxicol.* **32,** 377-397

10. Soderling, T.R. (1990) Protein kinases: Regulation by autoinhibitory domains. *J. Biol. Chem.* **265,** 1823-1826

11. Tokumitsu, H., Chijiwa, T., Hagiwara, M., Mizutani, A., Terasawa, M. & Hidaka, H. (1990) KN-62, 1-[N,O-bis(5-isoquinolinesulfonyl)-N-methyl-L-tyrosyl]-4-phenylpiperazine, a specific inhibitor of Ca^{2+}/calmodulin-dependent protein kinase II. *J. Biol. Chem.* **265,** 4315-4320

12. Ishii, A., Kiuchi, K., Kobayashi, R., Sumi, M., Hidaka, H. & Nagatsu, T. (1991) A selective Ca^{2+}/calmodulin-dependent protein kinase II inhibitor, KN-62, inhibits the enhanced phosphorylation and the activation of tyrosine hydroxylase by 56 mM K^+ in rat pheochromocytoma PC12h cells. *Biochem. Biophys. Res. Commun.* **156,** 1051-1056

13. Tsunoda, Y., Funasaka, M., Modlin, I.M., Hidaka, H., Fox, L.M. & Goldenring, R. (1992) An inhibitor of Ca^{2+}/calmodulin-dependent protein kinase II, KN-62, inhibits cholinergic-stimulated parietal cell secretion. *Am. J. Physiol.* **262,** G118-G122

14. Tohda, M., Nakamura, J., Hidaka, H. & Nomura, Y. (1991) Inhibitory effects of KN-62, a specific inhibitor of Ca/calmodulin-dependent protein kinase II, on serotonin-dependent protein kinase II, on serotonin-evoked Cl^- current and $36Cl^-$ efflux in *Xenopus* oocytes. *Neurosci. Lett.* **129,** 47-50

15. Ito, I., Hidaka, H. & Sugiyama, H. (1991) Effects of KN-62, a specific inhibitor of calcium/calmodulin-dependent protein kinase II, on long-term potentiation in the rat hippocampus. *Neurosci. Lett.* **121,** 119-121

16. Funauchi, M., Tsumoto, T., Nishigori, A., Yshimura Y. & Hidaka, H. (1992) Long-term depression is induced in Ca^{2+}/calmodulin kinase-inhibited neurones in visual cortex. *NeuroReport* **4,** 173-176

17. Tansey, M.G., Word, R.A., Hidaka, H., Singer, H.A., Schworer, C.M., Kamm, K.E. & Stull, J.T. (1992) Phosphorylation of myosin light chain kinase by the multifunctional calmodulin-dependent protein kinase II in smooth muscle cells. *J. Biol. Chem.* **267,** 12511-12516

18. Kamm, K.E. & Stull, J.T. (1989) *Annu. Rev. Physiol.* **51,** 299-313

19. Irons, C.E., Sei, C.A., Hidaka, H. & Glembotski, C.C. (1992) Protein kinase C and calmodulin kinase are required for endothelin-stimulated atrial natriuretic factor secretion from primary atrial myocytes. *J. Biol. Chem.* **267,** 5211-5216

20. Sumi, M., Kiuchi, K., Ishikawa, T., Ishii, A., Hagiwara, M., Nagatsu, T. & Hidaka, H. (1991) The newly synthesized selective Ca^{2+}/calmodulin dependent protein kinase II inhibitor KN-93 reduces dopamine contents in PC12h cells. *Biochem. Biophys. Res. Commun.* **181,** 968-975

21. Ito, M., Tanaka, T., Inagaki, M., Nakanishi, K. & Hidaka, H. (1986) N-(6-phenylhexyl)-5-chloro-1-naphthalenesulfonamide: A novel activator of protein kinase C. *Biochemistry* **25,** 4179-4184

22. Hidaka, H., Inagaki, M., Kawamoto, S. & Sasaki, Y. (1984) Isoquinolinesulfonamides, novel and potent inhibitors of cyclic nucleotide-dependent protein kinase and protein kinase C. *Biochemistry* **23,** 5036-5041

23. Chijiwa, T., Mishima, A., Hagiwara, M., Sano, M., Hayashi, K., Inoue, T., Naito, K., Toshioka, T. & Hidaka, H. (1990) Inhibition of forskolin-induced neurite outgrowth and protein phosphorylation by a newly synthesized selective inhibitor of cyclic AMP-dependent protein kinase, N-[2-(p-

bromocinnamylamino)ethyl-5-isoquinolinesufonamide (H-89), of PC12D pheochromocytoma cells. *J. Biol. Chem.* **265,** 5267-5272

24.   Thompson M.E., Zimmer, W.E., Wear, L.B., MacMillan, L.A., Thompson, W.J., Huttner, W.B., Hidaka, H. & Scammell, J.G. (1992) Differential regulation of chromogranin B/secretogranin I and secretogranin II by forskolin in PC12 cells. *Mol. Brain Res.* **12,** 195-202

25.   Chijiwa, T., Hagiwara, M. & Hidaka, H. (1989) A newly synthesized selective casein kinase I inhibitor, *N*-(2-aminoethyl)-5-chloroisoquinoline-8-sulfonamide, and affinity purification of casein kinase I from bovine testis. *J. Biol. Chem.* **264,** 4924-4927

7

Mitochondrial DNA and disease

Simon R. Hammans

Wessex Neurological Centre, Southampton General Hospital, Shirley, Southampton SO9 4XY, U.K.

The mitochondrion

The mitochondrion is a cellular organelle bound by a double membrane. It is approximately 0.5-1 μm in diameter, and may be spherical, elongated or branched. Typically, there are several hundred mitochondria per cell, but this varies with the respiratory activity of the tissue. The outer of the two mitochondrial membranes is permeable, allowing passage of solutes less than 10 kDa. The inner membrane encloses the matrix space, which is the domain of enzymes of the Krebs cycle and those involved in oxidation of pyruvate and fatty acids.

The primary function of the mitochondrion is to generate ATP, the energy currency of the cell. The enzymes of the respiratory chain and ATP synthase are assembled in the lipid bilayer of the inner mitochondrial membrane. The respiratory chain consists of four oligomeric peptide complexes linked by the carriers coenzyme Q_{10} and cytochrome *c* (Figure 1). Reducing equivalents generated by oxidation of pyruvate, fatty acids and the Krebs cycle are transferred to the respiratory chain via $NADH_2$ and $FADH_2$, and finally to molecular oxygen to form water. The chemiosmotic theory suggests that redox energy liberated at three sites (complexes I, III and IV) is coupled to generation of a proton gradient across the inner mitochondrial membrane, thus creating an electrical potential and a pH gradient. This electrochemical gradient is harnessed by ATP synthase (complex V) to drive ATP synthesis and also provides energy for translocation of ions and other metabolites across the inner mitochondrial membrane.

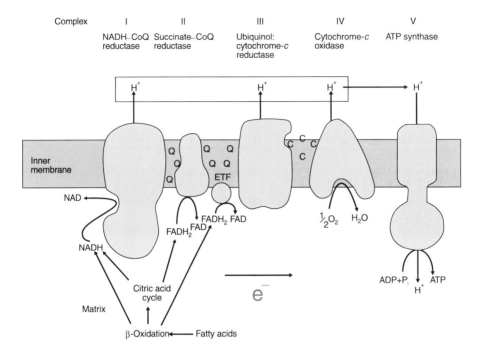

Figure 1. The mitochondrial respiratory chain
Q, coenzyme Q; C, cytochrome *c*; ETF, electron-transferring flavoprotein.

Mitochondrial DNA

Mitochondria are unique among mammalian cellular organelles in containing DNA. The mitochondrial genome is a closed, circular molecule of 16 569 base pairs (Figure 2), and consists of two strands, a G-rich heavy (H-) strand and C-rich light (L-) strand. It is estimated that 2-10 copies exist in each mitochondrion; mitochondrial DNA (mtDNA) contributes approximately 1% to total cellular DNA. The entire human mtDNA sequence has been determined[1].

MtDNA differs from nuclear DNA in its genetic code (Table 1), and also because it contains very little non-coding sequence[1]. Genes are frequently contiguous, occasionally overlap and do not contain introns. The only non-coding sequence is found in the D- (displacement) loop region of the genome, involved in initiation of mtDNA replication and transcription. MtDNA encodes its own translation apparatus; a complete set of 22 transfer (t) RNAs and 2 ribosomal (r) RNAs. Thirteen protein subunits are encoded; these provide an essential contribution to the respiratory chain complexes and ATP synthase (Table 2); the remaining subunits are encoded by nuclear DNA and imported into the mitochondrion.

MtDNA has been shown to be almost exclusively maternally inherited in mammals. The spermatozoon contains a few mitochondria located in the mid-piece behind the head, with about 50 mtDNA molecules in total. In contrast,

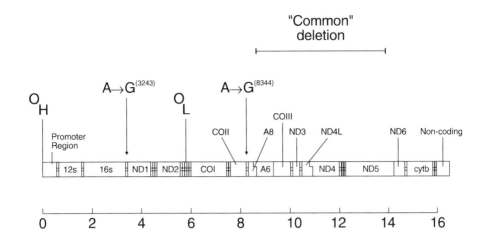

Figure 2. Linearized map of the mitochondrial genome
Scale in kb. Protein reading frames are labelled, tRNAs are shown as small boxes. The sites of the tRNA^{Lys} A→G$^{(8344)}$ mutation, the tRNA$^{Leu(UUR)}$ A→G$^{(8344)}$ mutation and the 'common' deletion are shown.

Table 1. The mitochondrial genetic code

First position	Second Position				Third position
	U	**C**	**A**	**G**	
U	Phe	Ser	Tyr	Cys	U
	Phe	Ser	Tyr	Cys	C
	Leu	Ser	Stop	Trp*	A
	Leu	Ser	Stop	Trp	G
C	Leu	Pro	His	Arg	U
	Leu	Pro	His	Arg	C
	Leu	Pro	Gln	Arg	A
	Leu	Pro	Gln	Arg	G
A	Ile	Thr	Asn	Ser	U
	Ile	Thr	Asn	Ser	C
	Met*	Thr	Lys	Stop*	A
	Met	Thr	Lys	Stop*	G
G	Val	Ala	Asp	Gly	U
	Val	Ala	Asp	Gly	C
	Val	Ala	Gln	Gly	A
	Val	Ala	Gln	Gly	G

*Differs from 'universal' (nuclear) code.

Table 2. Respiratory chain polypeptides

Complex		No. of subunits	No. encoded by mtDNA
I	NADH ubiquinone reductase (ND)	Up to 41	7 (ND1, ND2, ND3, ND4, ND4L, ND5, ND6)
II	Succinate dehydrogenase	4	0
III	Ubiquinone-cytochrome-c reductase	11	1 (cytochrome b)
IV	Cytochrome-c oxidase (COX)	13	3 (COX I, COX II, COX III)
V	ATP synthase (ATPase)	14	2 (subunits 6, 8)

the ovum contains much larger numbers of mitochondria. It is not clear whether the small paternal contribution to the zygote contributes significantly to mtDNA variation. During oogenesis, the number of mitochondria increases about 100-fold, while the number of mitochondrial genomes per organelle falls from 2 to 10 to about 1-2. After fertilization, the number of mitochondria and mtDNAs in the zygote does not change appreciably until the blastocyst stage, at which time mitochondrial division and mtDNA replication are believed to resume[2] and somatic cell levels of mitochondria and mtDNA are re-attained.

Sequence variation between maternal lines was recognized early in the study of mtDNA. No intra-individual variation has been recorded in normal human families, although this has been demonstrated in Holstein cows, *Drosophila* and some lower organisms. A series of studies showed that proportions of mutant and wild-type mtDNA molecules can shift during mitotic cellular replication by random partitioning of daughter cells during cytokinesis[3]. Lineages can drift towards purity of either mtDNA type or can remain heteroplasmic. By following the transmission of a heteroplasmic point mutation in a Holstein cow, Ashley *et al.*[4] were able to show marked shifts in heteroplasmy in a single generation, perhaps arising from the 'bottleneck' during oogenesis when the mitochondria selected may start with mtDNA of a pure genetic type.

Mitochondrial diseases

Mitochondrial myopathies
The term mitochondrial myopathy is applied to a clinically, biochemically and genetically heterogenous group of diseases which usually show mitochondrial structural abnormalities in skeletal muscle. The morphological hallmark of mitochondrial myopathies is the ragged red fibre (RRF) seen on transverse muscle sections stained by the modified Gomori trichrome (mGT) stain. The RRF represents an abnormal fibre segment with intermyobrillar and subsarcolemmal accumulations of mitochondria. Many ragged red segments show diminished or absent cytochrome-c oxidase (COX) activity when stained histochemically.

The first description of mitochondrial myopathy (MM) was in a patient with severe euthyroid hypermetabolism due to defective coupling of oxidation and phosphorylation in muscle mitochondria, with typical changes of mitochondrial myopathy on electron microscopy. In the 1970s, use of the mGT stain identified further cases of MM. Initially, patients with the syndromes of progressive external ophthalmoplegia (PEO) and/or muscle weakness often enhanced by exercise were observed. Subsequently patients with complex multisystem disorders predominantly affecting the CNS have been identified. These patients may have psychomotor retardation, dementia, ataxia, seizures, stroke-like episodes, myoclonus or other movement disorders, optic atrophy, pigmentary retinopathy, deafness or peripheral neuropathy in different combinations. Involvement of the heart, endocrine system, kidney, gut and haemopoietic tissues has also been reported.

Some, but not all, cases of MM form distinctive clinical syndromes:

- Kearns-Sayre syndrome (KSS) is defined as PEO and pigmentary retinopathy with onset earlier than the age of 20, with subsequent development of one of ataxia, heart block or raised CSF protein.
- Myoclonic epilepsy and ragged red fibres (MERRF), with the core features of myoclonic epilepsy and ataxia, with associated features of deafness, dementia and optic atrophy. The syndrome is often familial, and analysis of several pedigrees suggests maternal inheritance.
- Mitochondrial myopathy, encephalopathy, lactic acidosis and stroke-like episodes (MELAS), often associated with short stature. Transmission of MELAS occurs and is compatible with maternal inheritance.

Of patients with MM with similarly affected relatives, maternal transmission of MM to offspring is more frequent than paternal transmission, in a ratio of approximately 9:1. This provided a clue to the involvement of mtDNA in the aetiology of mitochondrial diseases. Over the last 5 years many different mtDNA defects have been described. These may be divided into those where the primary defect is thought to lie in nuclearly encoded genes, with secondary effects on mtDNA, or those with apparently primary defects of mtDNA (Table 3).

Mitochondrial DNA deletions

Holt and colleagues[5,6] established an association between mtDNA defects and human disease when they demonstrated that approximately 40% of patients with MM had two populations of mtDNA (heteroplasmy), one normal and one showing a large deletion up to 7 kb in size. The proportion of mutant mtDNA was between 20 and 70% in muscle, and lower in blood (usually undetectable by Southern blot). Almost all patients with single deletions have had PEO, often with one or more other features of KSS. Of patients with PEO, PEO 'plus' or KSS, 70-80% have heteroplasmic deletions of muscle mtDNA[6,7]. Duplications of mtDNA are rarer than deletions and may arise from deletions of mtDNA dimers.

Table 3. The involvement of mtDNA in the aetiology of mitochondrial diseases

Genome responsible	Class of mtDNA defect	Examples	Typical clinical features	Inheritance	RRF	Biochemical defect	Genetic defect detectable in blood?
Mitochondrial	Single heteroplasmic deletions	'Common' deletion (4.9 kb) accounts for 40% of cases	Nearly all have PEO +/− other features KSS Pearson's syndrome	Sporadic	Usually	Often generalized respiratory chain defect or normal	Usually at low level
Mitochondrial	Heteroplasmic duplications	7.6 kb duplication within COX I gene	KSS, diabetes mellitus, hypoparathyroidism	Sporadic	Yes		Yes
Mitochondrial	Heteroplasmic tRNA deletions	tRNA^{Lys} A → G$^{(8344)}$ tRNA$^{Leu(UUR)}$A → G$^{(3243)}$	MERRF MELAS Other encephalopathies PEO, KSS Myopathy Cardiomyopathy	Matrilineal	Usually	Often generalized	Yes

Table 3. Contd...

Genome responsible	Class of mtDNA defect	Examples	Typical clinical features	Inheritance	RRF	Biochemical defect	Genetic defect detectable in blood?
Mitochondrial	Protein coding genes	ND4 G→A(11778)	Leber's hereditary optic neuropathy, Muscle weakness, ataxia, retinopathy	Matrilineal, higher penetrance in males (see text)	No	? Complex I deficiency	Yes
		ATPase 6 T→G(8993)	Leigh's syndrome	Matrilineal	No	—	Yes
Nuclear	Multiple deletions		PEO plus other features, Metabolic crises, Myoglobinuria, Cardiomyopathy, Depression	Autosomal, usually dominant, some may be recessive	Usually	Generalized, predominantly COX deficiency	In some cases
Nuclear	MtDNA depletion	Organ specific	Renal tubulopathy, Hepatic failure, Myopathy	Probably autosomal recessive	Usually	COX deficiency	No

Pearson's marrow-pancreas syndrome presents in infancy; many patients die before the age of 3 years. Survivors have developed features of KSS in later childhood suggesting that Pearson's syndrome is part of the spectrum of MM. This is supported by the findings of large mtDNA deletions in muscle and blood, similar to those found in muscle in PEO and KSS[8].

Single heteroplasmic deletions are usually sporadic. In contrast, several syndromes of familial mitochondrial myopathy have been described in which genetic analysis reveals multiple different deletions. In some families, inheritance is clearly autosomal dominant. The deletions are thought to arise from defective mtDNA replication caused by a defect of one of the nuclearly encoded proteins involved in this process.

In most published biochemical studies of patients with mtDNA deletions, the rate-limiting step has been localized to complexes I, III, or IV, or combined deficiencies of complexes I-IV, III and IV[6,7]. A number of patients with deletions had no detectable biochemical defect of pooled mitochondria. The site of the deletion appears to have some bearing on the biochemical defect. A subgroup of patients with deletions confined to ND reading frames in addition to tRNA genes were found to have pure complex I deficiencies[6].

Pathogenesis of deletions

There are approximately 130 patients with mtDNA deletions described to date. About one-third of deletions are identical, being flanked by a 13 bp direct repeat existing at bp 8470-8482 and bp 13447-13459, and designated the 'common deletion'. The high incidence of this deletion may arise from the length of the direct repeat. In a study of 28 patients with 17 different mtDNA deletions[9], 12 had the common deletion and 8 had deletions flanked by other direct repeats 5-11 bp in length. The remaining 8 were flanked by imperfect repeats or by non-homologous sequences, suggesting that pathogenetic mechanisms may be heterogenous.

It is not clear when mtDNA deletions arise. Disease associated with deletions is usually sporadic, and most studies of mothers of patients have not found detectable levels of deletions. Deletions are widespread throughout different bodily tissues of affected patients. Consequently, deletions probably occur during oogenesis in most cases. A model involving slipped mispairing has been proposed. This requires that both direct repeats be present as complementary single strands, thus allowing pairing and recombination to occur. However, because normal mtDNA replication is an asynchronous and asymmetric process, repeats that are exposed at any one time will be direct rather than complementary, leading to criticisms of this model[9].

Patients with KSS and mtDNA deletions have shown widespread but uneven tissue distribution of deleted mtDNA, usually with only low levels detectable in blood. Variation in the proportion of mutant mtDNA between tissues possibly arises from random segregation of mtDNA and mitochondria in the oocyte or zygote, or more probably as a result of selection of mtDNA

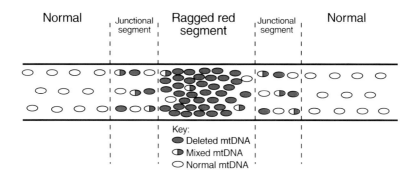

Figure 3. Schematic diagram showing the distribution of normal and deleted mtDNA in RRF segments in muscle of patients with mitochondrial myopathy

and mitochondria. Patterns of cell division, respiratory demands and nuclear gene expression differ between cell types, creating differing selection conditions for deleted and wild-type mtDNA.

In situ hybridization studies of muscle of patients with mtDNA deletions have elucidated the role of deleted mtDNAs in disease pathogenesis[10]. Deleted mtDNA appears to be confined to histochemically abnormal fibre segments and is transcribed, while normal mtDNA is depleted within these segments (Figure 3). Accumulations of mitochondria within RRF contain mostly deleted mtDNA with most organelles containing exclusively this species. Deletions of mtDNA create deficiencies of mRNA species derived from involved protein subunit reading frames, but all described large deletions also involve some tRNA genes. They would thus be expected to cause a translational defect involving all mtDNA-encoded subunits. Although the distribution of different mtDNA species between mitochondria is unclear, co-existence of deleted and normal mtDNA in the same mitochondrion could potentially ameliorate the translational defect, if normal mtDNA supplied missing tRNAs (complementation). A study of hybrid clones formed by fusing enucleated cultured fibroblasts containing deleted mtDNA with HeLa cells without mtDNA[11] showed that translational complementation between genomes occurred when the proportion of deleted mtDNA was less than 60%, while higher proportions were associated with progressive inhibition of translation and concomitant COX deficiency. Patients with deletions involving COX subunits tend to have absent COX activity in RRF, while COX activity is relatively preserved in RRF of patients with other deletions. Intramitochondrial, intergenomic complementation between deleted and normal mtDNA is the probable explanation of the histochemical differences between cases with or without involvement of COX subunits[12].

Transfer RNA point mutations

In 1990, Shoffner and colleagues[13] first described the association of an A→G[(8344)] mutation in mtDNA with the syndrome of MERRF. Disease severity shows some correlation with the proportion of mutant mtDNA in muscle or blood. The A→G[(8344)] mutation alters a conserved nucleotide in the TψC loop of the tRNA[Lys] gene. The TψC loop is thought to interact with the ribosomal surface, so that a mutation in this region is likely to affect incorporation of lysine residues into mitochondrial proteins. An *in situ* hybridization study has shown that abnormal fibre segments on muscle biopsy show an increased prevalence of mtDNA and a corresponding increase in mtRNA[10]. The biochemical findings in the proband of a large American pedigree were of combined deficiency of complexes I, IV and V, with reduced synthesis of larger mtDNA-encoded proteins in lymphoblastoid lines[14]; consistent with the hypothesis of tRNA[Lys] dysfunction.

An A→G[(3243)] mutation has been described in association with the syndrome of MELAS[15], and has subsequently been found in nearly 90% of MELAS probands. However, the mutation is also harboured by patients with other mitochondrial encephalopathies and a variety of other phenotypes including myopathies and diabetes and deafness. The mutation is heteroplasmic, maternally inherited and is found in blood.

The A→G[(3243)] mutation alters a conserved nucleotide in the dihydro-uridine loop of the tRNA[Leu(UUR)] gene. The mechanism of disease pathogenesis associated with this mutation remains unknown. In addition to affecting the function of tRNA[Leu(UUR)], it may interfere with H-strand transcription termination at the end of the 16S rRNA gene, mediated by the protein fraction mtTERM. MtTERM footprints a 13 bp sequence which includes the mutation site. Transcription termination has been shown to be inefficient in the presence of the mutation *in vitro*, which correlated with reduced affinity of the mtTERM fraction for the mutant sequence[16]. However, transfection studies[17] *in vitro* and *in situ* hybridization studies[10] have shown that there is no significant quantitative disturbance of H-strand transcription. It is unclear whether quantitative or qualitative changes in tRNA[Leu(UUR)], qualitative defects of transcription termination, or other abnormalities, are the dominant factors in pathogenesis.

MtDNA depletion syndrome

Moraes and colleagues[18] have described a fatal mitochondrial disease affecting different tissues (liver, muscle and kidney) in four different patients. Two infants were related, though not through the maternal line. *In situ* hybridization and biochemical evidence suggested an underlying tissue-specific mtDNA depletion. The pathogenesis is unclear, although a nuclear gene defect giving rise to defective replication of mtDNA was suggested.

Other mtDNA defects

Further, less common, mtDNA tRNA mutations have been described. A cluster of apparently pathogenic mutations exists in the tRNA$^{Leu(UUR)}$ gene. Known primary or secondary mtDNA defects now account for approximately 70% of patients with MM (A.E. Harding & S.R. Hammans, unpublished work). It is likely that further causative point mutations of mtDNA, and possibly nuclear DNA defects, will account for the remainder.

It is apparent that the known mtDNA defects underlying MMs (with their histological hallmark of RRF) may all affect tRNA function, either by deletion or point mutation, and thereby cause a defect of translation. To date, two syndromes associated with mutations of protein-encoding mtDNA genes have been described: Leber's hereditary optic neuropathy and a recently described syndrome associated with an ATPase mutation. Both are maternally inherited and, in contrast to MMs, are not associated with RRF.

Leber's hereditary optic neuropathy

Leber's hereditary optic neuropathy (LHON) is a disease which typically affects young adult males. Characteristically, both eyes are affected, with simultaneous or sequential involvement; visual loss is severe. Occasionally families show additional neurological features. The pattern of transmission of LHON suggests mitochondrial inheritance. Affected descendants of male patients (i.e. their daughters' sons) have never been described, making X-linked inheritance unlikely. About 85% of patients are male, and 18% of female carriers are affected. Between 50 and 100% of the sons of carriers are affected, and 70-100% of daughters of carriers are themselves carriers.

A point mutation of leucocyte mtDNA (at position 11 778) was reported in 9 of 11 LHON pedigrees in the USA, causing an amino acid change in subunit 4 of complex I of the respiratory chain[19]. In Finland and the U.K., 50-80% of LHON families have this mutation. More recently, two point mutations have been described in subunit 1 of complex I, and further point mutations have been described which appear to be pathogenic in combination, but not alone. The presence of a mtDNA mutation does not account for the excess of males with the disease, or the absence of disease in some family members with a high proportion of mutant DNA. A visual loss susceptibility locus on the X chromosome has been proposed. Such an interaction between nuclear and mitochondrial DNA would provide an explanation for the excess of males with the disease, since females would be required to be homozygous for the putative disease susceptibility allele. A study of British and Italian families has since excluded linkage at this locus, but linkage to other X-chromosome loci remains a possibility. Progress in defining the aetiology does not necessarily clarify the pathogenesis of LHON. The abrupt onset in early adult life and the apparent tissue specificity of the respiratory chain defect are unexplained. Environmental or immunological factors may provide an explanation.

Neurogenic muscle weakness, ataxia and retinitis pigmentosa

A family with these and other neurological features, but no RRF discernible on muscle biopsy, were heteroplasmic for a T→G[(8933)] transition in a subunit of ATPase (complex V). This mutation has also been observed in another family with similar clinical features. Two infants in this family died with a phenotype of Leigh's syndrome, suggesting that this mtDNA mutation may be one cause of this disorder.

Other putative mitochondrial diseases

There is some evidence to link Parkinson's disease with mitochondrial dysfunction. 1-Methyl-4-phenyl-1,2,3,6-tetrahydropyridine (MPTP) is a neurotoxin which induces parkinsonism in man. Its active metabolite, the 1-methyl-4-pyridinium ion (MPP[+]), is a complex I inhibitor. An anatomically selective deficiency of complex I has been reported in the substantia nigra of patients with Parkinson's disease[20], although more diffuse and/or widely distributed loss of respiratory chain function has been suggested by others in studies of muscle and platelets. Analysis of mtDNA from the brains of patients dying from Parkinson's disease has shown a low abundance of deleted mtDNAs detected by the polymerase chain reaction, but this was also the case in elderly normal subjects. The finding of low levels of mtDNA deletions in the normal elderly is of interest; the frequency of ragged red fibres and cytochrome oxidase deficiency in skeletal muscle also increases with age. It has been proposed that accumulation of mtDNA defects with age may contribute to the ageing process.

Multiple deletions of mtDNA at low levels have also been observed in three patients without neurological features who died from hypertrophic or dilated cardiomyopathies. The contribution of such genetic defects to the pathogenesis of cardiomyopathy is not clear.

The phenomenon of the phenotypic specificity of different mtDNA mutations raises difficult questions of how the genetic defect and resulting respiratory chain dysfunction can cause such disparate clinical manifestations. Necropsy studies argue against specific organ involvement being determined solely by uneven somatic distribution of the mutation. Clinical expression of a mtDNA defect may be modulated by other mtDNA or nuclear genes. Further specificity of disease expression may arise from the complex interaction of proteins within the mitochondrion, with diverse results in distinct cell types because of differential expression of genes and respiratory demands. In disorders associated with mtDNA defects, the complexity of pathogenesis is compounded not only by potential genetic-environmental interaction, but by unique interactions of two genomes.

Figure 1 was produced with the help of Dr Mark Cooper.

Suggestions for further reading

Revue Neurologique (1991) 147, 6-7 (Several articles in one issue)

Schapira, A.H.V. and DiMauro, S., (eds.) *Mitochondrial Disorders in Neurology*, Butterworth Heinemann, Oxford, in the press

References

1. Anderson, S., Bankier, A.T., Barrell, B.G., de Bruijn, M.H., Coulson, A.R., Drouin, J., Eperon, I.C., Nierlich, D.P., Roe, B.A., Sanger, F., Schreier, P.H., Smith, A.J., Staden, R. & Young, I.G. (1981) Sequence and organization of the human mitochondrial genome. *Nature (London)* **290**, 457-465

2. Piko, L. & Taylor, K.D. (1987) Amounts of mitochondrial DNA and abundance of some mitochondrial gene transcripts in early mouse embryos. *Dev. Biol.* **123**, 364-374

3. Wallace, D.C. (1986) Mitotic segregation of mitochondrial DNAs in human cell hybrids and expression of chloramphenicol resistance. *Somatic Cell Mol. Genetics* **12**, 41-49

4. Ashley, M.V., Laipis, P.J. & Hauswirth, W.W. (1989) Rapid segregation of heteroplasmic bovine mitochondria. *Nucleic Acids Res.* **17**, 7325-7331

5. Holt, I.J., Harding, A.E. & Morgan-Hughes, J.A. (1988) Deletions of muscle mitochondrial DNA in patients with mitochondrial myopathies. *Nature (London)* **331**, 717-719

6. Holt, I.J., Harding, A.E., Cooper, J.M., Schapira, A.H.V., Toscano, A., Clark, J.B. & Morgan-Hughes, J.A. (1989) Mitochondrial myopathies: clinical and biochemical features of 30 patients with major deletions of muscle mitochondrial DNA. *Ann. Neurol.* **26**, 699-708

7. Moraes, C.T., DiMauro, S., Zeviani, M., Lombes, A., Shanske, S., Miranda, A.F., Nakase, H., Bonilla, E., Werneck, L.C., Servidei, S., Nonaka, I., *et al.* (1989) Mitochondrial DNA deletions in progressive external ophthalmoplegia and Kearns-Sayre syndrome. *N. Engl. J. Med* **320**, 1293-1299

8. Rotig, A., Colonna, M., Blanche, S., Fischer, A., Le Deist, F., Frezal, J., Saudubray, J.M. & Munnich, A. (1988) Deletion of blood mitochondrial DNA in pancytopenia. *Lancet* **ii**, 567-568

9. Mita, S., Rizzuto, R., Moraes, C.T., Shanske, S., Arnaudo, E., Fabrizi, G.M., Koga, Y., DiMauro, S. & Schon, E.A. (1990) Recombination via flanking direct repeats is a major cause of large-scale deletions of human mitochondrial DNA. *Nucleic Acids Res.* **18**, 561-567

10. Hammans, S.R., Sweeney, M.G., Wicks, D.A.G., Morgan-Hughes, J.A. & Harding, A.E. (1992) A molecular genetic study of focal histochemical defects in mitochondrial encephalomyopathies. *Brain* **115**, 343-365

11. Hayashi, J., Ohta, S., Kikuchi, A., Takemitsu, M., Goto, Y. & Nonaka, I. (1991) Introduction of disease-related mitochondrial DNA deletions into HeLa cells lacking mitochondrial DNA results in mitochondrial dysfunction. *Proc. Natl. Acad. Sci. U.S.A.* **88**, 10614-10618

12. Hammans, S.R., Sweeney, M.G., Holt, I.J., Cooper, J.M., Toscano, A., Clark, J.B., Morgan-Hughes, J.A. & Harding, A.E. (1992) Evidence for intramitochondrial complementation between deleted and normal mitochondrial DNA in some patients with mitochondrial myopathy. *J. Neurol. Sci.* **107**, 87-92

13. Shoffner, J.M., Lott, M.T., Lezza, A.M., Seibel, P., Ballinger, S.W. & Wallace, D.C. (1990) Myoclonic epilepsy and ragged-red fiber disease (MERRF) is associated with a mitochondrial DNA tRNA(Lys) mutation. *Cell* **61**, 931-937

14. Wallace, D.C., Zheng, X., Lott, M.T., Shoffner, J.M., Hodge, J.A., Kelley, R.I., Epstein, C.M. & Hopkins, L.C. (1988) Familial mitochondrial encephalomyopathy (MERRF): genetic, pathophysiological, and biochemical characterization of a mitochondrial DNA disease. *Cell* **55**, 601-610

15. Goto, Y., Nonaka, I. & Horai, S. (1990) A mutation in the tRNA$^{Leu(UUR)}$ gene associated with the MELAS subgroup of mitochondrial encephalomyopathies. *Nature (London)* **348**, 651-653

16. Hess, J.F., Parisi, M.A., Bennett, J.L. & Clayton, D.A. (1991) Impairment of mitochondrial transcription termination by a point mutation associated with the MELAS subgroup of mitochondrial encephalomyopathies. *Nature (London)* **351**, 236-239

17. King, M.P., Koga, Y., Davidson, M. & Schon, E.A. (1992) Defects in mitochondrial protein synthesis and respiratory chain activity segregate with the tRNA[Leu(UUR)] mutation associated with mitochondrial encephalopathy, lactic acidosis, and strokelike episodes. *Mol. Cell. Biol.* **12,** 480-490

18. Moraes, C.T., Shanske, S., Tritschler, H.J., Aprille, J.R., Andreetta, F., Bonilla, E., Schon, E.A. & DiMauro, S. (1991) mtDNA depletion with variable tissue expression: a novel genetic abnormality in mitochondrial diseases. *Am. J. Human Genetics* **48,** 492-501

19. Wallace, D.C., Singh, G., Lott, M.T., Hodge, J.A., Schurr, T.G., Lezza, A.M., Elsas, L.J. & Nikoskelainen, E.K. (1988) Mitochondrial DNA mutation associated with Leber's hereditary optic neuropathy. *Science* **242,** 1427-1430

20. Schapira, A.H.V., Cooper, J.M., Dexter, D., Clark, J.B., Jenner, P. & Marsden, C.D. (1990) Mitochondrial complex I deficiency in Parkinson's disease. *J. Neurochem.* **54,** 823-827

8

PIG-tailed membrane proteins

Anthony J. Turner

Department of Biochemistry and Molecular Biology, University of Leeds, Leeds LS2 9JT, U.K.

Introduction

Cells make use of a variety of mechanisms to provide the stable association of proteins with membranes. In the majority of cases, it is one or more trans-membrane segments of the polypeptide chain itself that anchor the protein. An alternative mechanism incorporates a lipid anchor attached to the protein which can provide either a stable or transient interaction with the lipid bilayer. Often the lipid moiety provides the only contact between protein and membrane. The main focus of this essay will be one such form of lipid anchor which involves the covalent attachment to the protein of a glycosyl-phosphatidylinositol (GPI) glycolipid moiety. What began as a chance observation of the late 1970s relating to the release of a handful of cell-surface enzymes has developed rapidly into recognition of a common and functionally important mechanism for membrane association of eukaryotic proteins.

It is a diverse group of cell-surface proteins that is integrated into the plasma membrane by means of a GPI-anchor. The majority of protozoan parasitic cell-surface antigens are of this type. GPI-anchored mammalian proteins are less abundant, but examples range from complement regulatory factors, cell adhesion molecules, differentiation antigens, tumour markers, the prion protein, certain receptors, e.g. the folate receptor, and a collection of ectoenzymes. Table 1 lists some representative proteins that are established as GPI-anchored proteins, also sometimes referred to as 'glypiated' or 'PIG-tailed proteins' (PIG, phosphatidylinositol-glycan), since the anchor is always attached to the C-terminal 'tail' of the protein. Several of the proteins listed in Table 1 have contributed much to the GPI story, especially trypanosomal variant surface glycoprotein (VSG), the mammalian Thy-1 glycoprotein found in brain and the immune system, as well as the enzymes alkaline phosphatase and acetyl-cholinesterase. Some cell-surface proteins can occur either as a transmembrane protein or as a GPI-anchored form as a result of alternative splicing, a process that may be developmentally regulated. The neural cell-adhesion molecule (N-CAM) and acetylcholinesterase provide two specific examples of this phenomenon. The range of eukaryotic organisms that carry GPI-anchors is also highly diverse, ranging from a wide variety of mammalian cells and tissues through to yeasts and slime moulds although no such anchors have yet been reported in plants. GPIs can also occur in some organisms either unsubstituted or attached to polysaccharide rather than protein as in the parasitic protozoan *Leishmania.*

The nature of GPI anchors, especially of mammalian origin, their biosynthesis, and possible physiological functions, are addressed in this essay. More detailed consideration of aspects of GPI-anchored proteins are to be found in a

Table 1. Some examples of GPI-anchored membrane proteins

Enzymes

Alkaline phosphatase

Acetylcholinesterase

5′-Nucleotidase

Aminopeptidase P

Membrane dipeptidase

Trehalase

Promastigote surface protease (*Leishmania major*)

Cell-adhesion molecules

LFA-3 (lymphocytes)

N-CAM

Fibronectin receptors (subset)

Fasciclin I (*Drosophila*)

Mammalian antigens

Thy-1

Qa (mouse lymphocytes)

Ly-6 (mouse lymphocytes)

Carcino-embryonic antigen

Other

Variant surface glycoprotein (*T. brucei*)

Folate receptor

Ciliary neurotrophic factor (CNTF) receptor

Scrapie prion protein

Decay accelerating factor

Protectin

GP-2 (pancreatic zymogen granule)

Uromodulin (Tamm-Horsfall glycoprotein)

number of recent review articles and books[1-6]. The rapid development of this area owes much to the efforts of biochemists in fields as disparate as parasitology, neurobiology, haematology and carbohydrate chemistry. New insights have been provided into mechanisms of protein targeting, cell signalling and endocytosis.

Discovery of the GPI anchor

As so often in biology, the original observations that led to the identification of GPI anchors were entirely serendipitous. They arose from studies in the mid-1950s on anthrax-infected animals which were found to have levels of alkaline phosphatase in the blood some thirty times the normal value. The soluble factor secreted by *Bacillus anthracis* and responsible for the release of the enzyme was also secreted by some other bacterial species and shown to have phospholipase activity. More detailed studies showed that only phospholipases with specificity for phosphatidylinositol were able to cause release. Initial interpretations of the observations were that the phospholipase preparation may have been contaminated with proteinases, causing non-specific proteolysis of the protein from the membrane, or that the phospholipase had a disruptive effect on the cell membrane causing protein release. A series of careful studies of the phenomenon, particularly by Hiroh Izekawa and Martin Low and colleagues in the 1970s, demonstrated that only certain membrane proteins were releasable by bacterial phospholipases: these included 5′-nucleotidase in addition to alkaline phosphatase[1,7]. These studies led to the conclusion that these enzymes were attached to the plasma membrane by a covalent linkage between a site on the protein and the polar head-group of a phosphatidylinositol molecule in the lipid bilayer. Confirmation of this came from the direct demonstration of inositol in homogeneous preparations of human erythrocyte acetylcholinesterase by gas chromatography–mass spectrometry, and subsequently in other GPI-anchored proteins. Studies on the mammalian proteins were, however, limited by the relatively small quantities of these proteins available for chemical analysis of the anchor. A significant development came with the discovery that the surface coat protein of the trypanosome (*Trypanosoma brucei*) was anchored to the cell surface by a glycolipid structure which included phosphatidylinositol and, like the mammalian anchors, was releasable by phosphatidylinositol-specific phospholipase C (PI-PLC). The parasite contains as many as 10 million copies of the protein on its surface and therefore proved a valuable model system in which to study the biology of GPI anchors.

Structure of GPI anchors

The structures of trypanosome VSG, human erythrocyte acetylcholinesterase and rat brain Thy-1 glycoprotein were all solved in 1988[8-10]. Nuclear magnetic resonance spectroscopy, mass spectrometry and a series of enzymic and chemical digestions combined to provide the structures which are compared in Figure 1. What has emerged from these structural studies is the concept of a core structure for the anchor which consists of a tetrasaccharide of three mannose units and one glucosamine linked glycosidically to the 6-hydroxyl of phosphatidylinositol. The alkyl or acyl chains of the phospholipid provide the sole attachment of the protein to the plasma membrane. The terminal mannose unit links the anchor to the protein via a phosphodiester bond to ethanolamine which is, in turn, attached to the C-terminal amino acid of the mature polypeptide chain. This core structure is conserved from protozoa to humans implying a common biosynthetic pathway. There are, however, marked differences in the side-chains attached to the core glycolipid (Figure 1). For example, trypanosome VSG possesses a variable galactose side-chain of up to eight sugar residues. Mammalian anchors have an ethanolamine phosphate in place of the galactose side-chain and other variations are also observed.

There is also heterogeneity among GPI anchors in their lipid content. In the VSG from the bloodstream form of trypanosomes, the lipid is dimyristoyl glycerol, which arises by a process termed 'fatty acid remodelling', in which the more hydrophobic fatty acids initially present in the anchor are replaced sequentially with myristate by de-acylation and re-acylation[11]. In higher organisms the fatty acids are much more heterogeneous and commonly contain 1-alkyl-2-acyl glycerol. Another important modification seen in some mature GPI anchors is an additional acyl group (palmitate) directly attached to the inositol ring. This additional acylation renders the protein resistant to release from the membrane by PI-PLC and may serve a physiological role. Human erythrocyte acetylcholinesterase is so modified and the complete structure of the GPI anchor of this protein is shown in Figure 2.

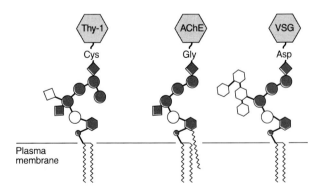

Figure 1. Outline structures of the glycolipid anchors of rat Thy-1, human acetyl-cholinesterase (AChE) and trypanosomal VSG
The C-terminal amino acid to which the anchor is attached is indicated. Note the differences in side-chains attached to the core glycolipid. The symbols represent: filled square, phospho-ethanolamine; filled circle, mannose; open circle, glucosamine; open square, N-acetylgalac-tosamine; open hexagon, galactose; filled hexagon, inositol and small filled circle, phosphate.

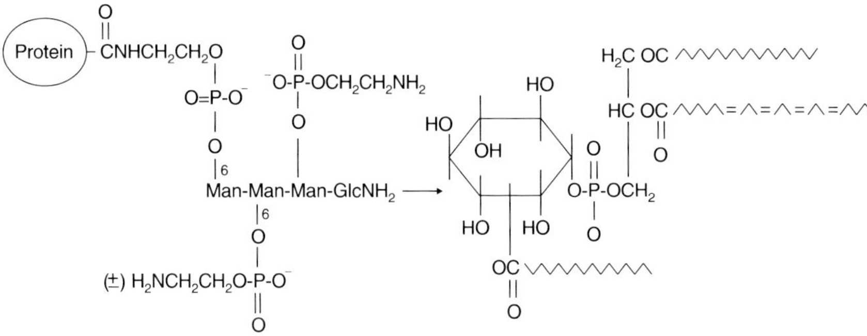

Figure 2. The structure of the glycolipid anchor of human erythrocyte acetyl-cholinesterase
Adapted from Deeg, et al. (1992) J. Biol. Chem. **267**, 18573-18580.

Signal sequence for GPI-anchor addition

As with other cell-surface proteins, those attached to the plasma membrane via a GPI anchor are initially synthesized on membrane-bound ribosomes and they carry an *N*-terminal signal peptide which is rapidly removed on translocation of the nascent protein into the lumen of the endoplasmic reticulum. For attachment of a GPI anchor, an additional cleavable hydrophobic signal is needed, this time located towards the *C*-terminus. The GPI anchor replaces the *C*-terminal signal in the mature protein.

Although the precise features of this signal remain to be defined, some aspects can be summarized. The mainly hydrophobic sequence at the *C*-terminus can be from 11 to 30 residues in length and has no cytoplasmic tail (Table 2). Site-directed mutagenesis has been used to determine key recognition elements for the transamidase enzyme that transfers the anchor to the protein. The precise *C*-terminal sequence is not critical, since it can be replaced by an unrelated sequence, provided it is generally hydrophobic in nature. The residue to which the anchor is attached is one of Gly, Asp, Asn, Ala, Ser or Cys, depending on the particular protein. This residue is usually separated from the hydrophobic sequence by a stretch of some 8-12 amino acids, mainly hydrophilic in character. The complete consensus signal for GPI-anchor attachment probably depends more on overall conformation and hydrophobicity rather than the linear amino acid sequence of this region of the protein. It is possible, however, to predict the site of addition of the anchor provided that the amino acid composition of the mature protein and the cDNA-derived amino acid sequence are known accurately[11a].

Biosynthesis of the GPI anchor

Kinetic studies indicate that the GPI anchor is added within one minute of polypeptide biosynthesis. This rapid transfer is supported by the observation

Table 2. Deduced C-terminal sequences of some GPI-anchored proteins

Protein	Species/tissue	C-terminal sequence
VSG (variant MIT 1170)	*T. brucei*	-NACK**D**SSILVTKKFALTVVSAAFVALLF
Promastigote surface protease	*L. major*	-KDGG**N**TAAGRRGPRAATALLVAALLAVAL
Thy-I	Rat	-KLVK**C**GGISLLVQNTSWLLLLLSLFTLQATDFISL
Alkaline phosphatase	Human placenta	-AGTT**D**AAHPGRSVVPALLPLLAGTLLLETATAP
5'-Nucleotidase	Human placenta	-RIKF**S**TGSHCHGSFSLIFLSLWAVIFVLYQ
Membrane dipeptidase	Human kidney	-YGYS**S**GASSLHRHWGLLLASLAPLVLCLSLL
Acetylcholinesterase	*Torpedo*	-NATA**C**DGELSSSGTSSSKGIIFYVLFSILYLIFY

The one-letter code for amino acids is used and the bold amino acids indicate the site of addition of the GPI anchor. Note the features of the C-terminal anchor: a small hydrophilic amino acid to which the GPI moiety is attached, a short stretch of mainly hydrophilic nature and then a predominantly hydrophobic sequence.

that a yeast mutant, defective in delivery of proteins from the endoplasmic reticulum to the Golgi apparatus, can still add a correct GPI anchor to a protein destined for the plasma membrane. This therefore suggests that the GPI anchors exist pre-assembled in the endoplasmic reticulum and, once synthesized, are quickly transferred as a complete unit to the primary translation products emerging from the ribosomes. Anchor addition probably involves the action of a transamidase enzyme which catalyses displacement of the C-terminal peptide by the amino group of the ethanolamine of the GPI precursor (Figure 3). The enzyme has not yet been isolated.

Cell-free systems have been used to study the biosynthesis of the GPI anchor precursor in yeast, trypanosomes and mammalian cells. In trypanosomes, the anchor is assembled by sequential glycosylation of phosphatidylinositol (PI). The first step involves the transfer of N-acetylglucosamine (GlcNAc) from UDP-GlcNAc to PI followed by de-acetylation of GlcNAc to glucosamine. This presence of a non-acetylated glucosamine is extremely unusual among biological glycoconjugates, but is found in all GPI anchors. Three gene products are required for the synthesis of GlcNAc-PI from UDP-GlcNAc and PI. Next, three mannose residues are added from the donor, dolichol-P-mannose, followed by addition of ethanolamine phosphate. The dolichol-P-mannose is probably synthesized on the cytoplasmic side of the membrane of the endoplasmic reticulum and subsequently translocated across to the lumenal side where the rest of the assembly takes place[12]. Fatty acid remodelling of the lipid moiety also appears to occur in the endoplasmic reticulum producing the glycolipid precursor termed 'glycolipid A'. The addi-

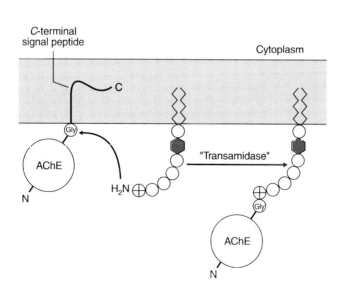

Figure 3. Model for the addition of a pre-formed glycolipid anchor to the tail of acetylcholinesterase (AChE) through the action of a putative transamidase

The C-terminal signal peptide of the protein is displaced by the amino group of the ethanolamine of the glycolipid precursor. The reaction occurs in the lumen of the endoplasmic reticulum.

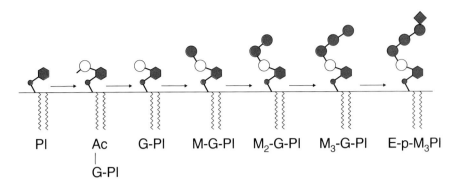

Figure 4. Outline of the biosynthesis of the core GPI anchor in protozoa by sequential glycosylation of PI

The symbols represent: phosphoethanolamine (E-p, filled square); mannose (M, filled circle); glucosamine (G, open circle); inositol (filled hexagon); phosphate (small filled circle); acetyl (Ac).

tion of side-chain galactose residues occurs later, probably in the Golgi apparatus. An outline of the initial stages of biosynthesis in the trypanosomal system are shown schematically in Figure 4. Both ATP and GTP appear to play a regulatory role in the biosynthesis of GPI-anchored proteins.

The biosynthetic pathways appear to be broadly similar in protozoa and mammals, although a few important differences do occur. These differences may be usefully exploited in anti-trypanosomal chemotherapy. In yeast and mammalian cells, the third reaction in the pathway appears to be the obligatory addition of an acyl chain to the inositol ring rendering the intermediate resistant to hydrolysis by PI-PLC. In most cases, this acyl group is removed at a much later stage in biosynthesis producing the final PI-PLC-sensitive GPI-anchored protein. The reasons for the transient acylation remain unclear[13]. Proteins which retain the acyl group in the mature protein, such as erythrocyte acetylcholinesterase, appear to be the exception rather than the rule. Another divergence in mammalian GPI anchor biosynthesis occurs after addition of the first mannose residue when additional phosphoethanolamine residues may be added. There are about ten genes encoding enzymes of the biosynthetic pathway, some of which have now been cloned[14], and we can expect further major advances in our understanding of the biochemistry of GPI anchor assembly in the coming years.

Paroxysmal nocturnal haemoglobinuria —- a defect in GPI anchor biosynthesis

Our knowledge of aspects of GPI anchor biosynthesis has evolved partly from studies of a rare blood disorder, paroxysmal nocturnal haemoglobinuria (PNH), which is an acquired form of haemolytic anaemia characterized by a population of blood cells that are abnormally sensitive to complement[15]. This hyper-sensitivity to complement arises from the lack of at least two GPI-anchored proteins on the surface of erythrocytes: decay-accelerating factor and protectin (membrane inhibitor of reactive lysis), which are membrane proteins that regulate the complement system and normally protect host tissues from autologous complement attack. Patients with PNH exhibit haemolysis of the abnormal red cells and hence haemoglobinuria, and have a tendency to thrombosis and, occasionally, a defect in the ability to fight infections. The accumulation of iron in the kidneys, as haemoglobin is catabolized in the proximal tubule and excreted, can even be in such a quantity as to set off airport metal detectors!

The disorder appears to arise from one or more defective steps in the GPI-anchor pathway in a bone marrow stem cell. Recent studies have tried to home in on the precise site of deficiency and studies with leucocytes from PNH patients suggest impaired glycosylation of the GPI-anchor occurs.

Functions of GPI anchors

The widespread occurrence of GPI anchors on proteins implies an important functional role besides the means to tether proteins at the cell surface. The occurrence of many mammalian GPI-anchored proteins in a soluble form in plasma suggests that their regulated release from the cell by specific phospholipases may be an important factor. It is noteworthy that plasma contains a GPI-specific phospholipase D activity that can cleave the GPI anchor although a physiological role for this enzyme has not been directly demonstrated. Table 3 lists some of the suggested functions for GPI anchors in eukaryotes. Some of these functions deserve more detailed discussion.

GPI anchors and cell signalling

Developments on two fronts have suggested that GPI-anchors may have a role to play in cell signalling, although this is currently a controversial subject. The first area is that of insulin signalling, specifically the possible role of second messengers in insulin action. Larner and colleagues[16] first reported the existence of an insulin-sensitive compound in skeletal muscle that could modulate the activity of glycogen synthase *in vitro*, and studies in other cell types have produced similar results, although the identity of the compounds involved has remained elusive. Initially thought to be peptide in nature, these enzyme modulators are now believed to resemble the glycolipid anchors and may, in some

---

Table 3. Some proposed functions of GPI-anchors

- Membrane anchorage
- Release from cell-surface by GPI-specific phospholipases C or D
- Cell adhesion and cell-cell interactions
- Increased membrane protein mobility
- Efficient membrane packing of proteins
- Protective role (e.g. VSG)
- Cell signalling
- Protein sorting and targeting
- Potocytosis

cases, be derived from them. Thus, Larner has proposed that 'insulin mediator' can be formed from GPI-linked protein precursors by a dual mechanism: both PLC and a putative protease are activated to produce diacylglycerol, mediator and release of the GPI-anchored protein. Internalization of the mediator is then required to allow it to act on intracellular enzymes hence regulating metabolism. It is still premature to regard these compounds as second messengers for any of the actions of insulin, or other hormones and growth factors, and several problems remain with the model. A primary requirement would be isolation and complete structural analysis of the putative mediator in pure form and demonstration that synthetic mediator can mimic some of the actions of insulin. The application of antibodies that recognize the PLC-cleaved GPI-anchor[17] might provide an alternative approach to elucidating the role of some of these molecules. For the moment, however, the precise mechanisms of action of insulin still remain elusive!

Signalling in the immune system has also implicated GPI anchors. Even before Thy-1 was known to possess a GPI anchor, it was shown that antibodies to the protein were mitogenic for T-cells. More than a dozen GPI-anchored membrane proteins have now been implicated in signalling in haemopoietic cells[18]. The GPI anchor seems to be directly involved in the signalling process. Comparison of GPI-anchored and transmembrane versions of a protein showed that only the GPI-anchored form is able to generate the signal. As with the insulin signalling, the mechanism of T-cell signalling by GPI anchors is problematic and the mechanism is unclear. Normal triggering of T-cells by antigen occurs through the T-cell receptor/CD3 complex that crosses the membrane and can generate intracellular signals. With T-cell activation through GPI anchors, the problem is that GPI-anchored proteins do not traverse the membrane, so how can an intracellular signal be generated? Plausible mechanisms include interaction of GPI-anchored proteins with signal-transducing proteins. In support of such a hypothesis is the ability of antibodies to certain T-cell GPI-anchored proteins to co-precipitate protein-tyrosine kinase activity. The precise nature of any interaction remains unclear, but recent evidence suggests that they may co-aggregate in large detergent-resistant domains in the cell membrane. It has long been known that GPI-anchored proteins exhibit insolubility in certain detergents, especially Triton X-100, and this can be used as a criterion for their identification[19]. The specific interaction of GPI-anchored proteins with specialized membrane domains enriched in sphingolipids and cholesterol may relate to the biological functions of the anchors[20], especially in cell signalling and in protein targeting (see below).

Protein sorting in polarized epithelial cells

In polarized epithelial cells specific mechanisms are required to sort and direct individual proteins to distinct regions of the plasma membrane. Various kidney cell lines have been used to demonstrate that GPI anchoring normally correlates with localization to the apical rather than the basolateral surface of the cell[21]. In hippocampal neurons, the GPI-anchored Thy-1 glycoprotein is directed exclusively to axonal membranes[22]. Thus the GPI anchor may represent a targeting signal and the segregation of this class of proteins into sphingolipid-rich microdomains in the trans-Golgi network may lead to their directed transport to the apical cell surface in 'apical transport vesicles' (Figure 5)[23]; a similar mechanism may operate in neuronal cells. Further support for the role of GPI anchors in targeting is the observation that the addition of a GPI tail to a protein that is normally present in the basolateral membrane redirects it to the apical membrane. However, apical localization of GPI-anchored proteins is not universal and other targeting signals are also important (see reference 20 for discussion of sorting mechanisms).

Potocytosis

Receptor-mediated endocytosis via clathrin-coated pits and phagocytosis are well-characterized pathways for the cellular uptake of macromolecules, such as the low-density lipoprotein (LDL) receptor, and for particulate material, respectively. Potocytosis is an alternative, clathrin-independent endocytic mechanism which can allow the concentration and uptake of small molecules and ions[24,25]. The observation that some vitamins, such as folate, can be delivered into cells by this mechanism has led to characterization of the uptake

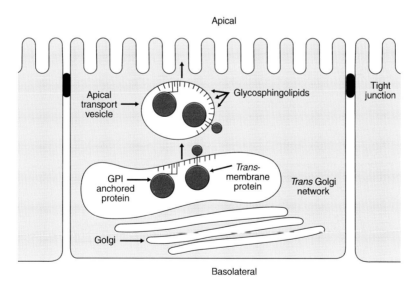

Figure 5. Schematic model for the sorting of apical membrane proteins in a polarized epithelial cell

The proteins are synthesized in the endoplasmic reticulum, transported through the Golgi stack and finally sorted into membrane vesicles in glycosphingolipid-rich membrane microdomains in the trans-Golgi network. The apical vesicles fuse with the apical domain to deliver their 'cargo'. GPI-anchoring normally correlates with apical localization whereas transmembrane proteins may be delivered to either the apical or the basolateral surface, depending upon additional unidentified sorting signals [adapted from Brown, D.A. (1992) *Trends Cell Biol.* **2**, 338-343].

pathway. The transport of 5-methyltetrahydrofolate into cells first involves its binding to a cell-surface folate receptor which is a GPI-anchored membrane protein. These receptors are clustered into regions of specialization of the plasma membrane referred to as caveolae which, in electron micrographs, appear as flask-shaped invaginations (Figure 6A). Other GPI-anchored proteins are also found to cluster in these regions. Binding of ligand induces the caveolae to close and pinch off from the plasma membrane, internalizing the receptor and its ligand (Figure 6B and C). Several hundred folate receptors are found in each caveola thus allowing substantial concentration of the vitamin against its concentration gradient. As with receptor-mediated endocytosis, low pH within the vesicle may be the trigger that induces the dissociation of folate from its receptor and its delivery into the cytoplasm. It has now proved possible to isolate caveolae biochemically, allowing their detailed structural analysis[26]. A major building block of the coat of the caveola is a 22 kDa protein named caveolin which presumably serves a similar role to clathrin in the conventional clathrin-coated pits.

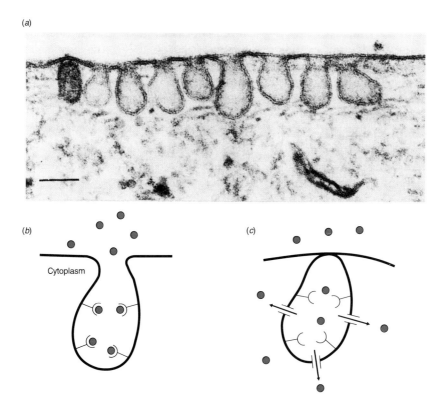

Figure 6. Illustration of receptor-mediated potocytosis involved in the uptake of, e.g. 5-methyltetrahydrofolate, into cells
(a) View of fibroblast caveolae by thin-section electron microscopy [from Rothberg *et al.* (1992) *Cell* **68**, 673-682]; (b) accumulation of ligand bound to GPI-anchored receptor in caveolae; (c) closure of caveolae and release of ligand into cytoplasm through change in ionic conditions (e.g. pH) within the caveolar compartment.

A number of the GPI-anchored proteins are ectoenzymes hydrolysing small molecules (peptides, nucleotides, oligosaccharides) (see Table 1) and enzyme-mediated potocytosis could allow the rapid hydrolysis of substrate and subsequent concentration of the products in caveolae facilitating their uptake. A further variant of this endocytic mechanism, carrier-mediated potocytosis, could allow the uptake of small hydrophobic molecules, e.g. the thyroid hormone thyroxine, bound to plasma transport proteins. The inositol-glycan 'insulin mediators' generated from GPI-anchored proteins may also be formed within caveolae and gain access to the cell interior by potocytosis.

Potocytosis therefore provides a versatile mechanism for concentrating and delivering small molecules and ions into the cell and this transport is critically dependent on interactions with GPI-anchored proteins. Receptor-mediated potocytosis can be exploited to allow the specific transport of foreign molecules such as therapeutic agents into cells. Even macromolecules linked to folate can be delivered into cells this way. Numerous medical and other applications can be envisaged for such an approach which may have advantages over other cell delivery systems such as permeabilization, microinjection or electroporation.

Other types of lipid anchor

While GPI-anchored proteins represent an important and rapidly expanding class of membrane proteins, other types of lipid anchors are also well recognized. None of these post-translational modifications represent a non-specific 'lipid glue' but they act to regulate a variety of cell functions. Other lipid modifications of proteins include the covalent attachment of long-chain acyl groups, particularly myristate and palmitate, and the modification by products of the mevalonic acid pathway: 15-carbon (farnesyl) or 20-carbon (geranylgeranyl) isoprenoid groups. Acylated proteins include some G-protein- and tyrosine kinase-linked receptors, G-protein subunits themselves, as well as some structural proteins, enzymes and viral proteins. Prenylated proteins include the oncogene product p21[ras] and other low molecular mass GTP-binding proteins, the nuclear lamins, the γ subunit of the G-proteins and various fungal mating factors. As with GPI anchors, such lipid modifications alter dramatically the hydrophobic character of proteins and facilitate their interaction with cell membranes. These modifications are likely to play important roles in the regulation of cellular functions and, since many of the proteins so modified are involved in cell signalling, may be important in cell growth and development. Other functions attributed to such modifications include protein targeting and transport, cytoskeletal organization, viral replication and protein folding and assembly. Understanding the biosynthetic pathways that lead to the addition of lipid anchors to proteins may lead to the development of novel agents with therapeutic potential in cancer, viral infections and in a range of other disease states.

Summary

- *Some membrane proteins are associated with the plasma membrane solely through a glycolipid moiety (GPI anchor).*
- *The GPI anchor is composed of a core structure of phosphatidylinositol attached to a glycan chain which, in turn, is attached to the C-terminus of the protein.*
- *The GPI-anchored protein can be released from the cell surface by the action of GPI-specific phospholipases C and D.*
- *In protozoa, GPI anchors represent the predominant mechanism for integrating cell-surface proteins into the lipid bilayer.*
- *Addition of a glycolipid anchor to a nascent protein requires a C-terminal hydrophobic signal sequence on the protein which is rapidly exchanged for a pre-assembled anchor.*
- *GPI anchors may have roles in protein targeting, cell signalling and in the uptake of small molecules (potocytosis).*
- *The human disease 'paroxysmal nocturnal haemoglobinuria' represents a defect in biosynthesis of the GPI anchor.*
- *Other lipid post-translational modifications of proteins are also recognized as important in regulating protein function (myristoylation, palmitoylation, prenylation).*

In the interests of brevity, I apologize to the many contributors to the field to whom I have not been able to refer and interested readers are directed to some of the more detailed reviews listed (references 1-6). I should like to thank the Medical Research Council and the Wellcome Trust for support of my research in this area and Nigel Hooper and other colleagues in Leeds, Mike Ferguson (Dundee) and Martin Low (New York) for fruitful collaborations and discussions over the years.

References

1. Low, M.G. (1989) The glycosyl-phosphatidylinositol anchor of membrane proteins. *Biochim. Biophys. Acta* **988**, 427-454
2. Doering, T.L., Masterson, W.J., Hart, G.W. & Englund, P.T. (1990) Biosynthesis of glycosyl phosphatidylinositol membrane anchors. *J. Biol. Chem.* **265**, 611-614
3. Turner, A.J. (ed.) (1990) The Molecular Biology of Membrane Proteins: Glycolipid Anchors of Cell-Surface Proteins, Ellis Horwood, Chichester
4. Cross, G.A.M. (1990) Glycolipid anchoring of plasma membrane proteins. *Annu. Rev. Cell Biol.* **6**, 1-39
5. Cardoso de Almeida, M.-.L (ed.) (1991) *Cell Biol. Int. Rep.* **15**, nos 9 and 11.
5a. McConville, M.J. & Ferguson, M.A.J. (1993) The structure, biosynthesis and function of glycosylated phosphatidylinositols. *Biochem. J.* **294**, 305-324.
6. Englund, P.T. (1993) The structure and biosynthesis of glycosyl-phosphatidylinositol protein anchors. *Annu. Rev. Biochem.* **62**, in the press
7. Low, M.G. & Finean, J.B. (1977) Release of alkaline phosphatase from membranes by a phosphatidylinositol-specific phospholipase C. *Biochem. J.* **167**, 281-284

8. Ferguson, M.A.J., Homans, S.W., Dwek, R.A. & Rademacher, T.W. (1988) Glycosyl-phosphatidyli-nositol moiety that anchors *Trypanosoma brucei* variant surface glycoprotein to the membrane. *Science* **239**, 753-759

9. Roberts, W.L., Santikarn, S., Reinhold, V. & Rosenberry, T.L. (1988) Structural characterization of the glycoinositol phospholipid membrane anchor of human erythrocyte acetylcholinesterase by fast atom bombardment mass spectrometry. *J. Biol. Chem.* **263**, 18776-18784

10. Homans, S.W., Ferguson, M.A.J., Dwek, R.A., Rademacher, T.W., Anand, R. & Williams, A.F. (1988) Complete structure of the glycosyl-phosphatidylinositol membrane anchor of rat brain Thy-1 glycoprotein. *Nature (London)* **333**, 269-272

11. Masterson, W., Doering, T.L., Hart, G.W. & Englund, P.T. (1989) A novel pathway for glycan assembly: biosynthesis of the glycosylphosphatidylinositol anchor of the trypanosome variant sur-face glycoprotein. *Cell* **56**, 793-800

11a. Antony, A.C. & Miller, M.E. (1994) Statistical prediction of the locus of the endoproteolytic cleav-age of the nascent polypeptide in glycosylphosphatidylinositol–anchored proteins. *Biochem J.* **298**, 9-16.

12. Abeijon, C. & Hirschberg, C.B. (1992) Topography of glycosylation reactions in the endoplasmic reticulum. *Trends Biochem. Sci.* **17**, 32-36

13. Field, M.C. (1992) Inositol acylation of glycosyl-phosphatidylinositol membrane anchors: what it is, and why it may be important. *Glycoconjugate J.* **9**, 155-159

14. Miyata, T., Takeda, J., Iida, Y., Yamada, N., Inoue, N., Takahashi, M., Maeda, K., Kitani, T. & Kinoshita, T. (1993) The cloning of PIG-A, a component in the early step of GPI-anchor biosyn-thesis. *Science* **259**, 1318-1320

15. Rosse, W.F. (1993) Evolution of clinical understanding: paraxysmal nocturnal hemoglobinuria as a paradigm. *Am. J. Hematol.* **42**, 122-126

16. Larner, J., Galasko, G., Cheng, K., DePaoli-Roach, A.A., Huang, L., Daggy, P. & Kellogg, J. (1979) Generation by insulin of a chemical mediator that controls protein phosphorylation and dephos-phorylation. *Science* **206**, 1408-1410

17. Hooper, N.M., Broomfield, S.J. & Turner, A.J. (1991) Characterization of antibodies to the glyco-syl-phosphatidylinositol membrane anchors of mammalian proteins. *Biochem. J.* **273**, 301-306

18. Robinson, P.J. (1991) Phosphatidylinositol membrane anchors and T cell activation. *Immunol. Today* **12**, 35-41

19. Hooper, N.M. & Turner, A.J. (1988) Ectoenzymes of the kidney microvillar membrane: differential solubilization by detergents can predict a glycosyl-phosphatidylinositol anchor. *Biochem. J.* **250**, 865-869

20. Brown, D.A. (1992) Interactions between GPI-anchored proteins and membrane lipids. *Trends Cell Biol.* **2**, 338-343

21. Lisanti, M.P. & Rodriguez-Boulan, E. (1990) Glycophospholipid membrane anchoring provides clues to the mechanism of protein sorting in polarized epithelial cells. *Trends Biochem. Sci.* **15**, 113-118

22. Dotti, C.G., Parton, R.G. & Simons, K. (1991) Polarized sorting of glypiated proteins in hippocam-pal neurons. *Nature (London)* **349**, 158-161

23. Simons, K. & van Meer, G. (1988) Lipid sorting in epithelial cells. *Biochemistry* **27**, 6197-6202

24. Anderson, R.G.W. (1993) Potocytosis of small molecules and ions by caveolae. *Trends Cell Biol.* **3**, 69-71

25. Hooper, N.M. (1992) More than just a membrane anchor. *Current Biol.* **2**, 617-619

26. Chang, W-J., Ying, Y-S., Rothberg, K.G., Hooper, N.M., Turner, A.J., Gambliel, H.A., de Gunzberg, J., Mumby, S.M., Gilman, A.G. & Anderson, R.G.W. (1994) Purification and characterization of smooth muscle cell caveolae. *J. Cell Biol.* in the press

Horseradish peroxidase: the analyst's friend

Orlaith Ryan, Malcolm R. Smyth* and Ciarán Ó Fágáin

*School of Biological Sciences and *School of Chemical Sciences, Dublin City University, Dublin 9, Republic of Ireland*

Introduction

Horseradish peroxidase (HRP) has long been a very powerful and useful scientific tool. It is one of the most widely used indicator enzymes in the life sciences. This is due to its high catalytic rates on a variety of substrates, its molecular stability and its ease of coupling to carriers or to other functional molecules such as antibodies. Here we survey the properties of HRP that fit it so well for analytical applications. We also outline the more prominent of these, including its use in biosensors.

The HRP enzyme

HRP is an oxidoreductase (donor: hydrogen-peroxide oxidoreductase; EC 1.11.1.7). It is a haemoprotein that transfers hydrogen from hydrogen donors to H_2O_2, as do all peroxidases[1]. HRP isoenzyme C is a single glycosylated polypeptide chain. The native enzyme consists of a haemin prosthetic group and 308 amino acid residues, including four disulphide bridges and two Ca^{2+} ions. These properties and the amino acid sequence of HRP isoenzyme C were elucidated by Welinder[2]. The relative molecular mass of native HRP is 44 000. One mole of protohaemin IX per HRP molecule acts as a prosthetic group. This protoporphyrin IX group is held in place by electrostatic interactions between the propionic side-chain of the haem and a lysine molecule in the apoprotein. The covalent structure of HRP isoenzyme C consists of two com-

pact domains, between which the haemin group is positioned[2]. The iron group
has six co-ordinate positions, of which four are occupied by porphyrin nitro-
gen atoms and the fifth by a protein group. The sixth can be occupied by various compounds. Peroxidases appear to operate by exchange of substrate in this
6th position[1].

HRP has carbohydrate side-chains attached to asparagine residues[2] at
eight different sites. Carbohydrates are more abundant in the *C*-terminal
region of the molecule. The characteristic absorption spectrum of native peroxidase shows a major Soret band at 403 nm, as depicted in Figure 1. The green
Compound I (see below) has absorption maxima at 407 and 658 nm, while the
pink Compound II shows maximal absorbances at 417, 530 and 561 nm[3]. The
enzyme contains a single tryptophan residue which fluoresces but is not located in the same domain as the active site. HRP activity is measured indirectly
by the rate of transformation of the hydrogen donor. The pH range of HRP
activity is 4.0-8.0[3]. Specific formation of compound I, as shown in Figure 2,
can be obtained only with certain peroxides. The specific hydrogen acceptors
that can be used are hydrogen-, methyl- and ethyl peroxide. Only these three
hydrogen acceptors are active, whereas a wide range of hydrogen donors
react[1]. These include a large number of phenols, aminophenols, indophenols,
diamines and leuco dyes. The reaction of various hydrogen donors and other
substrates with HRP will be discussed below ('HRP-catalysed reactions').

Catalysis

The reaction scheme for peroxidase catalysis is shown in Figure 2. Peroxidase
combines with hydrogen peroxide (substrate) which oxidizes its haem group
to form Compound I. In the absence of suitable electron donors, or at low
peroxide concentrations, Compound I decomposes slowly[4]. In the presence of
one-electron donors (including most of the chromophores used in peroxidase
assays), Compound I is reduced in sequential, one-electron transfer steps. The

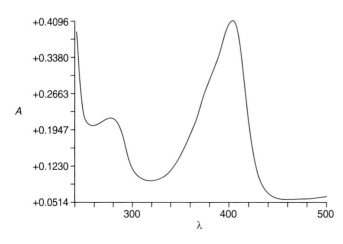

**Figure 1.
Absorption
spectrum of native
horseradish
peroxidase**
Note the characteris-
tic peak or Soret band
at 403 nm in addition
to the u.v. absorbance.

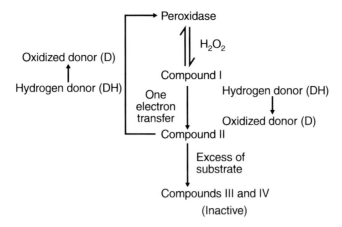

Figure 2. Scheme of reaction of HRP
Compounds I and II are formed during the course of the catalytic cycle. If excess substrate is present, Compound II may not regenerate free peroxidase, but may form the inactive Compounds III and IV (see reference 1).

first of these yields Compound II and the free radical resulting from one-electron oxidation of the donor. The second regenerates the native enzyme and the oxidized donor. Therefore, the original state is restored by two successive reductive steps of Compound I by a hydrogen donor. Most assays (spectrophotometric) measure the appearance of oxidized donor. (See 'HRP-catalysed reactions' below.) The structure of the enzyme-substrate compounds of peroxidase have been determined by X-ray absorption spectroscopy[5]. Valence states of the various compounds were also determined. The mechanism was elucidated by analysis of the structural changes observed in the reaction intermediates. A similar study on the oxidation-reduction potentials of the compound intermediates in the HRP reaction was undertaken by Hayashi & Yamazaki, who attempted to identify the structure of the iron atom of Compound II[6]. Techniques used to study the compounds included nuclear magnetic resonance (n.m.r.), electron spin resonance (e.s.r.) and magnetic circular dichroism (c.d.). A red or green colour (Compounds III and IV, respectively) will appear if there is substrate inhibition. There is a narrow optimum concentration range for hydrogen peroxide. This lies roughly between 0.1 and 2 mM depending on the hydrogen donor being used. If too low a concentration of hydrogen peroxide is used, there is decreased activity, whereas too high a peroxide concentration brings about substrate inhibition. An excess of hydrogen peroxide will inactivate the enzyme. Cyanide and sulphide reversibly inhibit HRP, but carbon monoxide does not. HRP is quite sensitive to bacteria, bacteriostatic agents and other chemicals found in tap water. It is recommended that ultrapure water be used to wash all glassware and in preparation of all reagents to be used in connection with HRP[1]. HRP may be inactivated by polystyrene plates if Tween 20 is omitted from buffers[7]. HRP's reaction kinetics have been reviewed in detail by Dunford[8] and it is also capable of reacting with various chemical forms of iodine[9].

Stability properties

Thermal inactivation kinetics of HRP have been studied in some detail. HRP is one of the most heat-stable enzymes in vegetables and has high inherent thermostability. Peroxidase causes unfavourable colour and off-flavour during food preservation processes[10]. The level of peroxidase activity remaining indicates the effectiveness of blanching treatments in food processing and this is one reason why the heat inactivation of HRP is of interest. Chang *et al.* examined the thermal inactivation kinetics of HRP in the presence of sugars by differential scanning calorimetry (d.s.c.) and also by determination of residual peroxidase activity[10]. Generally, a deviation from first-order kinetics is seen for the residual decay curve of HRP. This has been explained by several mechanisms, including the formation of enzyme aggregates with different heat stabilities, the presence of heat-stable and heat-labile isoenzymes and series-type enzyme inactivation kinetics. Chang and co-workers confirmed that HRP thermo-inactivation does not follow first-order kinetics. Both d.s.c. and residual activity determinations showed that the apparent reaction order for thermo-inactivation was 1.5. The presence of isoenzymes of different thermal resistance was thought to be responsible for the observed kinetics. Electrophoresis and isoelectric focusing revealed seven or more proteins from the original 'peroxidase' preparation. Four of these were peroxidase isoenzymes. The 1.5 order of thermal inactivation could then be accounted for by the presence of heat-stable and heat-labile isoenzymes, i.e. as a result of the heterogeneity of the HRP solution studied. The effect of sucrose, a non-reducing sugar, on the thermal stability of peroxidase was studied by d.s.c. Addition of increasing concentrations of sucrose resulted in an increase in the maximum denaturation temperature, indicating that sucrose addition stabilized the enzyme against thermal denaturation. At low sucrose concentrations, as revealed by the determination of residual activity, there was in fact a decrease in thermal stability of HRP. Reducing sugars such as fructose, glucose and lactose incubated with HRP brought about a more rapid inactivation of HRP than did sucrose[10]. Hendrickx *et al.* found solid-state, lyophilized HRP to be much more thermostable than HRP in solution: inactivation temperatures were in the range 140–160°C. Again the thermal inactivation curve was biphasic[11].

Ugarova *et al.* investigated the thermostability of HRP that had been chemically modified at the ε-amino groups of lysines with carboxylic acid anhydrides and picryl sulphonic acid. An increase in the thermostability of the modified enzyme was due to a decrease in conformational mobility in the protein moiety around the haem. It was the degree of modification, i.e. the number of modified ε-amino groups of lysines, rather than the nature of the modifier that was important. Modification of the enzyme resulted in restricted conformational mobility, as seen from c.d. spectra. The reduced flexibility, seen at four of the six lysines of HRP, was correlated to enhanced stability. However, modification of all six lysines, which resulted in reduced conformational flexibility of the protein in the vicinity of the haem, caused reduced thermostability[12].

Techniques such as enzyme immobilization, protein engineering and other physical and chemical modification procedures can affect the thermostability of an enzyme. Weng et al.[13] studied the thermal stability of immobilized peroxidase and compared the stability with native, soluble peroxidase[13]. Inactivation curves for soluble peroxidase showed biphasic behaviour. There are two theories to explain this. The first is the two fraction theory in which a heat-labile and a heat-stable fraction exist, as described by Chang et al.[10]. The second theory puts forward the idea that a partially denatured intermediate is formed during the heating process. This intermediate, or partially denatured — but active — peroxidase, has a higher thermal stability than the native enzyme. With peroxidase that is covalently immobilized on glass beads, biphasic behaviour was observed for temperatures below 80°C. At temperatures above 80°C, the inactivation followed first-order kinetics. The heat inactivation of peroxidase immobilized on glass beads has also been studied in a variety of organic solvents. This is of particular interest since, as discussed later, the enzyme may be used in organic solvents in a variety of applications.

HRP is a metalloprotein in which calcium contributes to the structural stability of the protein by maintaining its molecular conformation[14]. Bound calcium may be removed by incubation of the peroxidase with guanidine hydrochloride and EDTA. Calcium removal decreases the thermal stability of the enzyme compared with the native enzyme. This indicates that the Ca^{2+} ions function in maintaining the conformation of the enzyme. The function of calcium in a number of metalloproteins is to stabilize the enzyme structure.

Recombinant HRP

A synthetic gene encoding HRP isoenzyme C has been synthesized and expressed in *Escherichia coli*[15]. The gene that was constructed was based on the amino acid sequence of the mature protein, as described by Welinder[2]. A non-glycosylated recombinant enzyme was produced in an insoluble inactive form. (Insoluble products frequently result when heterologous proteins are expressed in *E. coli*.) The HRP C produced was solubilized and active enzyme was obtained when specific folding conditions were used. The reduced, denaturant-solubilized polypeptide was folded in the presence of specific concentrations of urea, Ca^{2+} and haem to give active enzyme. However, purification of active recombinant HRP C yielded about half the activity of native HRP C, when assayed under similar conditions. It thus appears that glycosylation is not essential for correct folding and activity. Since the *N*- and *C*-termini are not encoded in the construction of the gene, neither terminus is required for enzyme activity. Ca^{2+}, as discussed by Haschke & Friedhoff[14], is important in the folding and structure of the enzyme. The binding of Ca^{2+} is an obligatory step in folding before correct disulphide bridge formation and haem incorporation can be completed. Peroxidase is the only haemoprotein in which Ca^{2+} have been reported as a constituent. In the future, site-directed mutagenesis may be used to identify Ca^{2+}-binding regions. Site-directed mutagenesis has already been used to probe the catalytic mechanism, concentrating on residues close to the distal histidine in position 42[16]. (Most haemoproteins contain both a close or proximal histidine, which co-ordinates with the fifth position of the iron atom held in the haematin ring, and a more distant or distal histidine which participates in catalysis after binding of substrate.) Phe-41 was converted to Val (F41V) and Trp (F41W). Both F41V and F41W had diminished k_{cat} values compared with the 'wild-type' recombinant isoenzyme C using 2,2′-azino-di-[3-ethyl benzothiazoline-sulphonate] (ABTS) as substrate under steady-state conditions. These and other changes were thought to be due to a perturbation in the HRP substrate-binding site. Substitution of Arg-38 by Lys (R38K) led to decreased catalytic rates and to a 100-fold increase in the apparent K_m for H_2O_2, confirming the importance of this conserved Arg residue in HRP catalysis. It is hoped that the use of non-glycosylated recombinant HRP C may yield crystals suitable for X-ray crystallographic analysis so that the three-dimensional structure of HRP may be determined. Fujiyama and colleagues have isolated, cloned and characterized three cDNAs and two genomic DNAs corresponding to the mRNAs and genes for HRP isoenzyme C. Both genes consisted of four exons and three introns. An apparent promoter sequence was observed 5′ upstream, while a poly(A) signal occurred 3′ downstream in both genes[17].

HRP-catalysed reactions

Peroxidases are enzymes that catalyse the oxidation of various substrates. HRP decomposes two molecules of hydrogen peroxide, the natural substrate, into water and oxygen by a two-electron oxidation step. Thus, hydrogen peroxide is reduced in the presence of a hydrogen donor. The specificity that HRP has for the second molecule of hydrogen peroxide is low and many other electron donors may be used[18]. The native enzyme is regenerated by electron transfer from a hydrogen donor, which is oxidized. This is a redox or oxidation/reduction type reaction[19]. Monitoring of the oxidized donor gives an indirect means of monitoring the concentration of hydrogen peroxide involved in the reaction.

The detection and quantitative determination of hydrogen peroxide is of importance in industrial and clinical applications. The presence of peroxidase activity is widely used as a detection step in immunoassays, histochemical and immunoblotting procedures. Because of the widespread use of HRP and the wide need for the detection of H_2O_2, a large number of procedures have been developed for these purposes. Hydrogen peroxide is the natural substrate, but indicator molecules, which can be used to monitor activity, are often referred

Table 1. Common colorimetric substrates for HPR

* Refers to colours and wavelengths used when the action of HRP is terminated with acid. For full details of these and other substrates, see reference 1.

Compound	Carcinogen/mutagen	Colour	$\lambda_{max.}$(nm)
ABTS	Yes	Green	405
OPD	Yes	Orange	450
			492*
TMB	No	Blue	370,650
		Yellow*	450*

to as substrates. Substrates used to demonstrate the presence of peroxidase activity are numerous, widely available and many are well characterized. Substrates that are involved in colorimetric, fluorescent, chemiluminescent, electrochemical and hydroxylation reactions are all used to detect HRP activity, and to quantify hydrogen peroxide. All these techniques have relative merits.

Colorimetric and fluorimetric assays

Many chromogenic substrates act as hydrogen donors that form a coloured product on oxidation. The appearance of the product can be monitored spectrophotometrically, often using a microplate reader. Chromogenic substrates for HRP include o-phenylene diamine (OPD), ABTS and 3,3′,5,5′-tetramethylbenzidine (TMB)[20] (see Table 1). 4-Chloro-1-naphthol (4-CN), 3,3′-diaminobenzidine (DAB) and 3-methyl-2-benzothiazoline hydrazone hydrochloride (MBTH) are also used[18]. Chromogenic peroxidase substrates are often divided into two categories, depending on whether they form soluble or insoluble products. Insoluble product formation can be valuable for histochemical work, whereas soluble products are required for direct absorbance measurements. A chromogenic compound used in a detection step should be relatively inexpensive, easy to use, soluble, non-carcinogenic and otherwise safe to handle. The product should also be safe, stable and have high molar absorptivity at a convenient wavelength. Many HRP chromogens (including ABTS, DAB and OPD) are mutagenic or carcinogenic[1] and others lack sensitivity. MBTH and TMB do not appear to have any mutagenic or carcinogenic effects[1,34]. HRP can also use fluorescent substrates such as homovanillic acid (4-hydroxy-3-methoxy-phenylacetic acid). Their use, however, demands the availability of a fluorimeter, a specialized and costly item of equipment.

Chemiluminescent assays

Chemiluminescence is another reaction type catalysed by HRP. Chemiluminescence refers to the emission of light occurring during a chemical

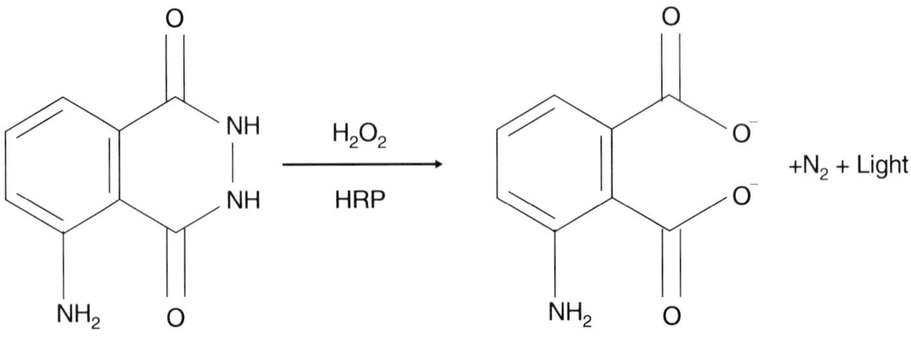

Figure 3. Reaction of luminol with HRP to yield 3-aminophthalate and light
For a more detailed treatment, see reference 42.

reaction. The most commonly used chemiluminescent reagents are luminol and related hydrazides[21]. HRP can oxidize luminol in an alkaline solution in the presence of hydrogen peroxide. The products of the reaction are 3-aminophthalate and light[22] (see Figure 3). In coupled enzyme assays, the amount of hydrogen peroxide that is produced by an oxidase may be assayed by using excess HRP and luminol in the reaction mixture. The steady-state chemiluminescence intensity can be measured at 425 nm. Hydrogen peroxide can be detected at very low levels using this system and this has been applied to the analysis of several enzymes and metabolites of clinical interest[23]. Immobilized luminol chemiluminescence reagent systems can be used to determine hydrogen peroxide in flowing stream/bioreactor situations. Enhancers such as luciferin, p-iodophenol or p-hydroxycinnamic acid are used to increase light emission from a flash signal to a steady-state output.

Electrochemical reactions

Electrochemical reactions involving HRP utilize a hydrogen donor that can be monitored voltametrically upon reduction. The reduction of the oxidized donor can be monitored and, therefore, the concentration of hydrogen peroxide involved in the reaction can be determined. A hydrogen donor that can act as an electron mediator and that can be followed voltametrically upon reduction is essential[24]. Electron mediators that act as hydrogen donors useful in the peroxidase-catalysed reduction of H_2O_2 include hydroquinone, o-toluidine, resorcinol and catechol (see Figure 4). Mediator/redox electrodes employing HRP will be discussed under 'Biosensors' below.

Other reactions of HRP

HRP can function as a catalyst in a number of water-immiscible organic solvents. Many enzymes can act as catalysts in nearly anhydrous organic solvents. This is because the water essential for enzymic activity is tightly bound to the

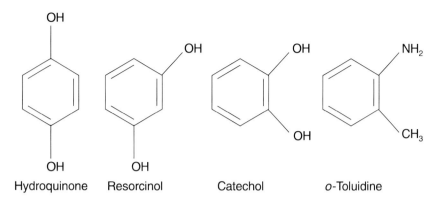

Hydroquinone   Resorcinol   Catechol   *o*-Toluidine

Figure 4. Structures of some hydrogen donors useful in HRP-mediated electrochem-ical reactions

enzyme molecules and may remain bound even when the bulk water is replaced by organic solvents. Enzymic function in organic solvents allows the determination and quantification of analytes that are not soluble in water/buffer systems[25]. These include cholesterol, which can be determined by a bi-enzyme system involving cholesterol oxidase and HRP.

The reactions catalysed by HRP in organic solvents include hydroxyla-tions, *N*-demethylations, sulphoxidations and other oxidations of various organic substances[26]. Chemical modification of HRP can render the enzyme more active and soluble in organic solvents. Donors used for HRP catalysis in conjunction with hydrogen peroxide include *p*-anisidine and 9-oxoellipticine.

Peroxidases can catalyse one-electron oxidations of phenols. Phenols that can be acted on by HRP include *p*-hydroxyphenylacetate and *p*-cresol. This results in the formation of phenoxy radicals, which spontaneously couple to form polymers[27]. The enzyme-catalysed phenoxy radical polymerization is favoured over other polymerization reactions used to oxidize phenols. Lignin is a phenolic resin and *p*-cresol can be incorporated into this via a peroxidase-catalysed co-polymerization in non-aqueous media. A simplified reaction scheme is depicted in Figure 5.

HRP can catalyse the hydroxylation of some aromatic compounds by molecular oxygen in the presence of dihydroxyfumaric acid as the hydrogen donor[28]. HRP can catalyse the oxidation of 9-methoxyellipticine to 9-oxoellip-ticine in the presence of H_2O_2. This reaction can be performed in diethylether if an enzyme modified with dithioesters is used[29]. See Figure 6 for examples of these reactions.

Applications of HRP

HRP is widely used in enzyme assays, immunoassays, immunohistochemical procedures and in bi-enzyme systems which generate hydrogen peroxide. Tijssen details the properties of HRP that make it suitable as an enzyme label

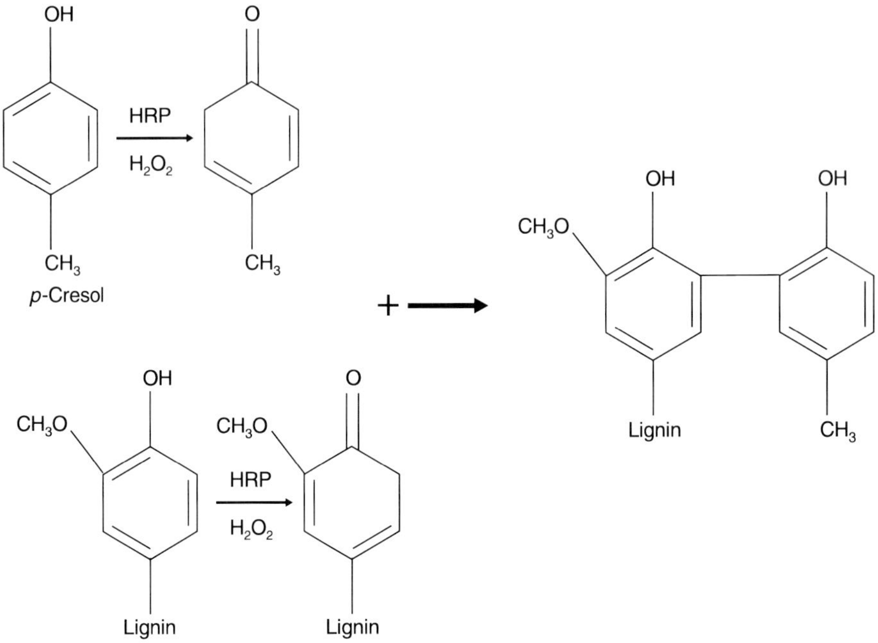

Figure 5. Simplified reaction scheme for HRP-mediated incorporation of *p*-cresol into lignin
See reference 25 for details.

in enzyme-linked immunosorbent assays (e.l.i.s.a.)[1]. These include high turnover number, stability upon storage and when used under various different assay conditions, i.e. variations in pH, ionic strength, buffer types and temperatures. There are few HRP inhibitors and these are not usually present as interfering substances in samples to be assayed, such as blood and urine. Also, HRP is reasonably cheap and is available in relatively pure form. It is well suited for the preparation of enzyme-conjugated antibodies owing to its excellent stability characteristics and ability to yield chromogenic products. As discussed above, its activity is easily detected using a wide range of substrates by colorimetric, fluorimetric and luminescent procedures, resulting in highly specific and sensitive assays. For enzyme immunoassays (e.i.a.), spectrophotometric assays using colorimetric substrates for the HRP enzyme label are generally employed. Closely related analytes can be distinguished by antibodies and it is this fact that is exploited in e.i.a. Antibodies can be used to quantify the amount of antigen in a sample, with a high degree of accuracy. One type of immunoassay is e.l.i.s.a. The antibody of interest is immobilized on to a solid support; the sample is added; unbound sample is removed by washing; then the second antibody (specific for a different site on the antigen and labelled with an enzyme) is added. Less than a nanogram of protein can be detected by the rapid and convenient e.l.i.s.a. The amount of second antibody that binds is

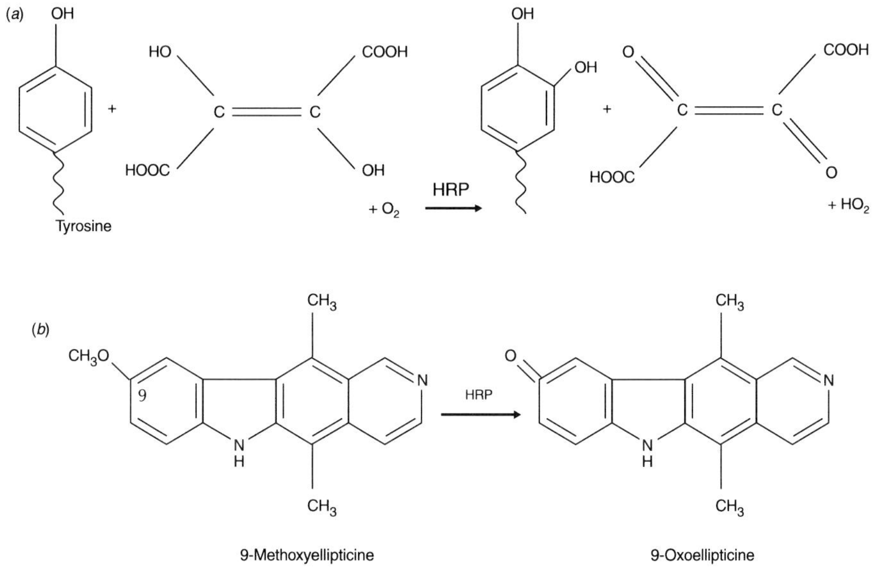

(a)

Tyrosine

+ O$_2$ →(HRP)

+ HO$_2$

(b)

CH$_3$O

9

9-Methoxyellipticine

→(HRP)

9-Oxoellipticine

Figure 6. Hydroxylation and oxidation reactions of HRP
(a) Hydroxylation of tyrosine to L-DOPA by HRP with dihydroxyfumaric acid as hydrogen donor. See reference 28 for this and further examples. (b) Oxidation of 9-methoxyellipticine to 9-oxoellipticine by HRP. See reference 43 for a detailed treatment.

proportional to the quantity of antigen of interest in the sample. The enzyme-second antibody conjugate can convert an added colourless substrate into a coloured product or a non-fluorescent substrate into a fluorescent product. The amount of product formed will be proportional to the amount of antigen. This type of sandwich immunoassay is a widely used technique. Its principle is shown in Figure 7.

Conjugation of antibody to enzyme label must be easily accomplished and the enzyme-antibody conjugates produced must be active and stable. Nakane & Kawaoi used sodium periodate to conjugate HRP to antibodies[30]. Here the carbohydrate moiety on HRP is oxidized by periodate. The resulting aldehyde groups form Schiff bases with non-protonated amino groups of the antibody. This method has been modified and improved so that highly active conjugates are obtained with high recovery[31]. HRP contains six lysine groups and these are often targeted in conjugation procedures using cross-linking reagents. The use of heterobifunctional cross-linkers, i.e. containing two different reactive groups, is a preferred method of conjugate production as the protein and antibody of interest may be reacted in a step-wise manner. This reduces the occurrence of unwanted side-reactions. Nilsson *et al.* conjugated HRP and immunoglobulins using a heterobifunctional reagent, *N*-succinimidyl 3-(2-pyridyldithio) propionate (SPDP)[32]. Its reaction scheme is shown in Figure 8. Many other conjugation procedures exist for the preparation of enzyme-antibody conjugates[1,33].

HRP is also used in immunohistochemical and immunoblotting proce-

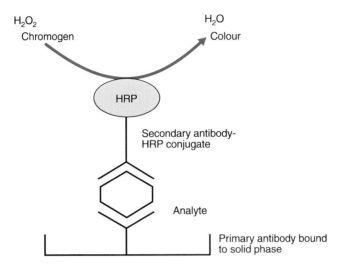

Figure 7. Principle of sandwich immunoassay
Primary antibody is immobilized on a solid phase. The analyte will bind to the antibody on addition of sample. After a washing step to remove unbound sample, a second, peroxidase-conjugated antibody is added and this binds to another region of the analyte. Unbound conjugate is removed by washing and the presence of bound conjugate is detected by the addition of a suitable substrate mixture. The intensity of the signal is proportional to the amount of analyte.

dures. Enzyme-labelled reagents used in immunohistochemical and immunoblotting procedures are detected using soluble chromogenic substrates which precipitate after enzyme action. When they precipitate, these substrates leave an insoluble coloured product at the site of bound enzyme. Antibodies can be raised to specific molecules, for example, proteins from brain or other organs. The exact functional location of a particular molecule *in vivo* can be located and visualized by this procedure of immunolocalization. HRP neurohistochemistry is one of the most frequent methods used for tracing neuronal connectivity within the central nervous system[34]. This technique involves the tracing of neural connections after the injection of HRP[35]. HRP has proven valuable in demonstrating uptake and retrograde axonal transport of exogeneous proteins by neurons[36]. In this procedure, HRP is injected into the test specimen. After a specific period of time, tissue slices from the brains of specimens are fixed. A chromogen that yields insoluble granules is used to stain for peroxidase activity. A study of the sites at which HRP activity is visualized can be used to give details about what nerves of the cortex, and other parts of the brain, transport HRP and, therefore, proteins. TMB is often used as substrate in these histochemical applications. Another use of HRP in physiological studies is in the study of fluid pinocytosis[37], the uptake of media by enclosure in small membrane vesicles that bud from the cell surface. This process, either of fluid or adsorptive pinocytosis, occurs in nearly all cells. Fluid pinocytosis can be observed and quantified by using HRP as a marker, since it is readily solu-

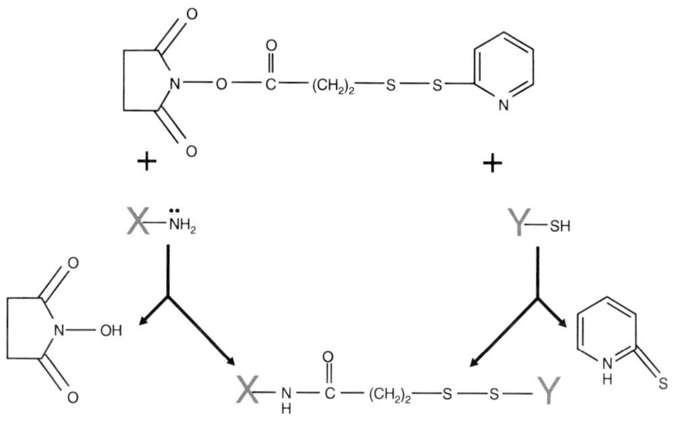

Figure 8. Structure and reaction of the heterobifunctional cross-linker SPDP
X and Y are two different proteins. For a full discussion, see reference 32.

ble, membrane-impermeable and does not alter cellular activities. Also, HRP is not metabolized within the cell, has no binding affinity for the plasma membrane and the formation of insoluble reaction products can be readily visualized microscopically.

HRP has been successfully used as a non-isotopic label in DNA probes. The enzyme is modified with p-benzoquinone and polyethyleneimine. This coats the HRP with an excess of positive charges to allow it to bind to negatively charged DNA. This derivatized HRP is then conjugated to the single-stranded oligonucleotide probe by glutaraldehyde cross-linking. Figure 9 shows the general scheme of reaction. The HRP-labelled probe is allowed to hybridize with the single-stranded DNA of interest[38]. The presence of the target sequence is indicated by HRP activity, as detected by a colorimetric or luminescent substrate. This system has been commercialized[39].

Bi-enzyme systems often incorporate the use of HRP. The schematic representation of a bi-enzyme system using HRP is shown in Figure 10. The use of such bi-enzyme systems is incorporated in many bioassays that do not directly produce a detectable product. Immobilization of these systems on to an electrode is frequently used in biosensors. HRP is often used to convert the H_2O_2 product from an oxidase reaction into a detectable form.

Biosensors

A biosensor comprises a biochemically responsive material immobilized in close proximity to a suitable transducing element, designed to convert a biochemical response into an electrical response. The biocomponent may be an enzyme, lectin, antibody or antigen, micro-organism, liposome, receptor, organelle or whole cell. The essence of a biosensor is two transducers. These two transducers relate the concentration of an analyte to a measurable electri-

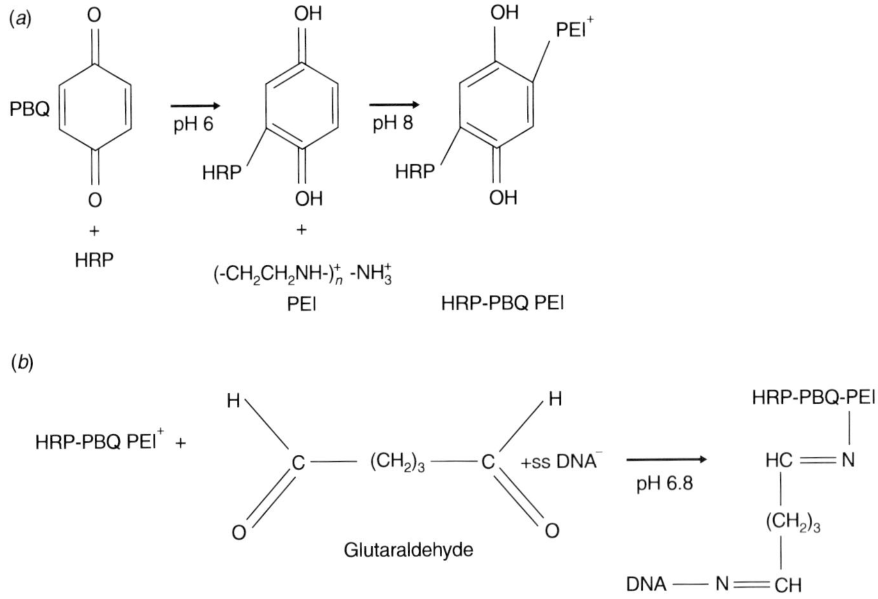

(a)

(b)

Figure 9. Labelling of HRP with PBQ/PEI and its conjugation to a DNA probe
PBQ (p-benzoquinone) reacts with functional groups on the HRP protein in the first step of a two-stage reaction. This is followed by reaction with polyethylenimine (PEI) (a). The resulting HRP derivative is positively charged and will bind to negatively charged single-stranded DNA (ssDNA). The two molecules are then covalently joined by glutaraldehyde cross-linking (b). For further details, see references 38,39.

cal signal. The biochemical transducer converts the analyte into a chemical and/or physical response, which is detected and converted into an electrical signal by the physical transducer[40]. The strength of the signal is related to the concentration of the analyte. A schematic diagram of a biosensor is shown in Figure 11. Using such devices a wide variety of analytes may be measured.

Bi-enzyme systems that incorporate HRP are frequently used in electro-chemical sensors. Substrates that can be monitored voltametrically are used

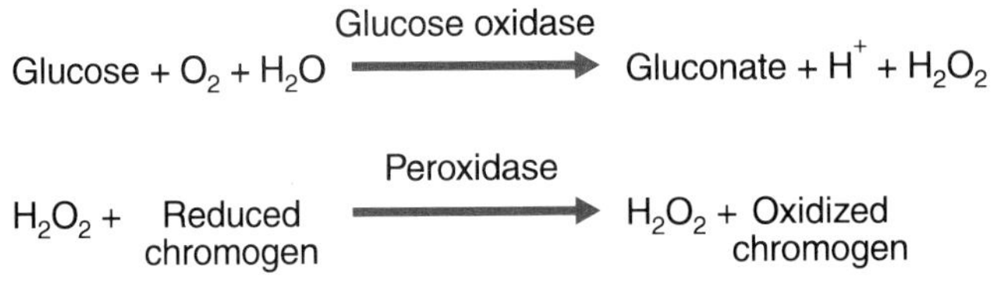

$$\text{Glucose} + O_2 + H_2O \xrightarrow{\text{Glucose oxidase}} \text{Gluconate} + H^+ + H_2O_2$$

$$H_2O_2 + \begin{array}{c}\text{Reduced}\\\text{chromogen}\end{array} \xrightarrow{\text{Peroxidase}} H_2O_2 + \begin{array}{c}\text{Oxidized}\\\text{chromogen}\end{array}$$

Figure 10. A bi-enzyme system involving HRP
The hydrogen peroxide produced in the glucose oxidase reaction acts as substrate for HRP which, by oxidizing the chromogen, allows easy quantification of glucose.

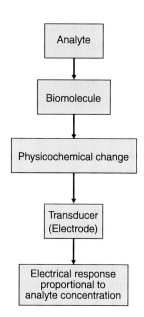

Figure 11. Schematic diagram of a biosensor
Note that the biomolecule may also be considered a transducer, since it converts one type of signal (biochemical) into another (physicochemical).

(see 'Electrochemical reactions' above). The first enzyme in a bi-enzyme system is often an oxidase that produces hydrogen peroxide.

A bi-enzyme sensor for the determination of alcohol, D-amino acid, L-amino acid, choline and cholesterol has been prepared using peroxidase and the respective oxidases[41]. The electrodes prepared contained the enzymes immobilized on graphite electrodes previously modified with TTF (tetrathiofulvalene) and TCNQ (tetrathiafulvalinium tetracyanoquinodimethanide). These redox compounds allow the graphite electrode to respond to the enzyme-substrate reaction. The mediator in the system was ferrocyanide.

A sensitive electrochemical assay for low levels of hydrogen peroxide was described by Frew *et al.*[19]. The system was based on the enzymic reduction of H_2O_2 by peroxidase and subsequent electron transfer from a gold or pyrrole graphite electrode to the enzyme, via a redox mediator. Recently, Blum has described a biosensor for the determination of glucose in drinks using a flow-injection analysis system based on co-immobilized glucose oxidase and HRP. HRP's chemiluminescent reaction with luminol is used in the fibre-optic detection system. The detection limit for glucose was 0.25 nmol. Up to 25 samples could be processed in 60 min and results agreed well with a standard hexokinase-based spectrophotometric method[44].

Conclusion

HRP can be used in a wide range of applications in analytical, industrial and clinical situations. A range of assays is available to monitor HRP activity. Because of its stability, ready availability and ease of use, it is likely that HRP will continue to act as an important detector enzyme in increasingly sophisticated systems and devices in the future.

Summary

- *HRP is a calcium-containing, extensively glycosylated, stable haemoprotein possessing four disulphide bridges and weighing 44 kDa.*

- *It is an oxidoreductase which can use a wide variety of hydrogen donors to reduce hydrogen peroxide. This gives rise to a range of colorimetric, fluorimetric, chemiluminescent and electrochemical assays for HRP activity.*

- *HRP is very widely used as an indicator in immunoassays, non-isotopic DNA probes, cytochemistry, bi-enzyme systems and biosensors. This is owing to its stability, high catalytic rates, ease of conjugation to other molecules and wide choice of assays of activity.*

Orlaith Ryan received financial support from Dublin City University and from Eolas, the Irish Government's science and technology agency. Robert O'Connor prepared many of the Figures and Enda Miland determined the HRP spectrum.

References

1. Tijssen, P. (1985) Practice and theory of enzyme immunoassays, in *Laboratory Techniques in Biochemistry and Molecular Biology* (Burdon, R.H. and van Knippenberg, P.H. eds., vol. 15, pp. 173-189, 231-239 and 249-261, Elsevier, Amsterdam

2. Welinder, K.G. (1979) Amino acid sequence studies of horseradish peroxidase. Amino and carboxyl termini, cyanogen bromide and tryptic fragments, the complete sequence, and some structural characteristics of horseradish peroxidase C. *Eur. J. Biochem.* **96**, 483-502

3. Paul, K.G. (1963) Peroxidases, in *The Enzymes* (Boyer, P.D., ed.), vol. 8, pp. 227-237, Academic Press

4. Pruitt, K.M., Kamau, D.N., Miller, K., Mansson-Rahemtulla, B. & Rahemtulla, F. (1990) Quantitative, standardized assays for determining the concentrations of bovine lactoperoxidase, human salivary peroxidase, and human myeloperoxidase. *Anal. Biochem.* **191**, 278-286

5. Chance, B., Powers, L., Ching, Y., Poulos, T., Schonbaum, G.R., Yamazaki, I. & Paul, K.G. (1984) X-Ray absorption studies of intermediates in peroxidase activity. *Arch. Biochem. Biophys.* **235**, 596-611

6. Hayashi, Y. & Yamazaki, I. (1979) The oxidation-reduction potentials of compound I/compound II/ferric couples of horseradish peroxidase A_2 and C*. *J. Biol. Chem.* **254**, 9101-9106

7. Berkowitz, D.B. & Webert, D.W. (1981) The inactivation of horseradish peroxidase by a polystyrene surface. *J. Immunol. Methods* **47**, 121-124

8. Dunford, H.B. (1991) Horseradish Peroxidase: structure and kinetic properties, in *Peroxidases in Chemistry and Biology* (Everse, J., Everse, K.E. & Grisham, M.B., eds.), vol. II, Chap. 1, pp. 1-24 CRC Press, Boca Raton FL, U.S.A.

9. Bhattacharyya, D.K., Bandyopadhyay, U., Chatterjee, R. & Banerjee, R.K. (1993) Iodide modulation of the EDTA-induced iodine reductase activity of horseradish peroxidase by interaction at or near the EDTA-binding site *Biochem. J.* **289** 575-580

10. Chang, B.S., Park, K.H. & Lund, D.B. (1988) Thermal inactivation kinetics of horseradish peroxidase. *J. Food Sci.* **53**, 920-923

11. Hendrickx, M., Saraiva, J., Lyssens, J., Oliveira, J. & Tobback, P. (1992) The influence of water activity on thermal stability of horseradish peroxidase *Int. J. Food Sci. Technol.* **27** 33-40

12. Ugarova, N.N., Rozhkova, G.D. & Berezin, I.V. (1979) Chemical modification of the ε-amino groups of lysine residues in horseradish peroxidase and its effect on the catalytic properties and thermostability of the enzyme. *Biochim. Biophys. Acta* **570**, 31-42

13. Weng, Z., Hendrickx, M., Maesmans, G., Gebruers, G. & Tobback, P. (1991) Thermostability of soluble and immobilized horseradish peroxidase. *J. Food Sci.* **56**, 574-578

14. Haschke, R.H. & Friedhoff, J.M. (1978) Calcium-related properties of horseradish peroxidase. *Biochem. Biophys. Res. Commun.* **80**, 1039-1042

15. Smith, A.T., Santama, N., Dacey, S., Edwards, M., Bray, R.C., Thorneley, R.N.F. & Burke, J.F. (1990) Expression of a synthetic gene for horseradish peroxidase C in *Escherichia coli* and folding and activation of the recombinant enzyme with Ca^{2+} and heme. *J. Biol. Chem.* **265**, 13335-13343

16. Smith, A.T., Sanders, S.A., Greschik, H., Thorneley, R.N.F., Burke, J.F. & Bray, R.C. (1992) Probing the mechanism of horseradish peroxidase by site-directed mutagenesis *Biochem. Soc. Transact.* **20**, 340-345

17. Fujiyama, K., Takemura, H., Shibayama, S., Kobayashi, K., Choi, J.-K., Shinmyo, A., Takano, M., Yamada, Y. & Okada, H. (1988) *Eur. J. Biochem.* **173**, 681-687

18. Conyers, S.M. & Kidwell, D.A. (1991) Chromogenic substrates for horseradish peroxidase. *Anal. Biochem.* **192**, 207-211

19. Frew, J.E., Harmer, M.A. Hill, H.A.O. & Libor, S.I. (1986) A method for estimation of hydrogen peroxide based on mediated electron transfer reactions of peroxidases at electrodes. *J. Electroanal. Chem.* **201**, 1-10

20. Madersbacher, S. & Berger, P. (1991) Double wavelength measurement of 3,3',5,5'-tetramethyl-benzidine (TMB) provides a three-fold enhancement of the ELISA measuring range. *J. Immunol. Methods* **138**, 125-128

21. Coulet, P.R. & Blum, L.J. (1992) Bioluminescence/chemiluminescence based sensors. *Trends Anal. Chem.* **11**, 57-61

22. Hool, K. & Nieman, T.A. (1988) Immobilized luminol chemiluminescence reagent system for hydrogen peroxide determinations in flowing streams. *Anal. Chem.* **60**, 834-837

23. Roda, A., Girotti, S., Grigolo, B., Ghini, S., Carrea, G., Bovara, R., Zini, I. & Grimaldi, R. (1991) Microdialysis and luminescent probe: Analytical and clinical aspects. *Biosensors Bioelectron.* **6**, 21-29

24. Sanchez, P.D., Blanco, P.T., Fernandez Alvarez, J.M., Smyth, M.R. & O'Kennedy, R. (1990) Flow-injection analysis of hydrogen peroxide using a horseradish peroxidase-modified electrode detection system. *Electroanalysis* **2**, 303-308

25. Kazandjian, R.Z., Dordick, J.S. & Klibanov, A.M. (1986) Enzymatic analyses in organic solvents. *Biotech. Bioeng.* **XXVIII**, 417-421

26. Urrutigoity, M. & Souppe, J. (1989) Biocatalysis in organic solvents with a polymer-bound horseradish peroxidase. *Biocatalysis* **2**, 145-149

27. Popp, J.L., Kirk, T.K. & Dordick, J.S. (1991) Incorporation of *p*-cresol into lignins via peroxidase-catalysed copolymerization in nonaqueous media. *Enzyme Microb. Technol.* **13**, 964-968

28. Klibanov, A.M., Berman, Z. & Alberti, B.N. (1981) Preparative hydroxylation of aromatic compounds catalyzed by peroxidase. *J. Am. Chem. Soc.* **103**, 6263-6264

29. Souppe, J., Urrutigoity, M. & Levesque, G. (1988) Application of the reaction of dithioesters with ε-amino groups in lysine to the chemical modification of proteins. *Biochim. Biophys. Acta* **957**, 254-257

30. Nakane, P.K. & Kawaoi, A. (1974) Peroxidase-labelled antibody: a new method of conjugation. *J. Histochem. Cytochem.* **22** 1084-1091

31. Tijssen, P. & Kurstak, E. (1984) Highly efficient and simple methods for the preparation of peroxidase and active-peroxidase conjugates for enzyme immunoassays. *Anal. Biochem.* **136**, 451-457

32. Nilsson, P., Bergquist, N.R. & Grundy, M.S. (1981) A technique for preparing defined conjugates of horseradish peroxidase and immunoglobulin. *J. Immunol. Methods* **41**, 81-93

33. Peeters, J.M., Hazendonk, T.G., Beuvery, E.C. & Tesser, G.I. (1989) Comparison of four bifunctional reagents for coupling peptides to proteins and the effect of the three moieties on the immunoassay of the conjugates. *J. Immunol. Methods* **120**, 133-143

34. Mesulam, M.M. (1978) Tetramethyl benzidine for horseradish peroxidase neurohistochemistry: A non-carcinogenic blue reaction-product with superior sensitivity for visualizing neural afferents and efferents. *J. Histochem. Cytochem.* **26**, 106-117

35. Olucha, F., Martinez-Garcia, F. & Lopez-Garcia, C. (1985) A new stabilizing agent for the tetramethylbenzidine (TMB) reaction product in the histochemical detection of horseradish peroxidase (HRP). *J. Neurosci. Methods.* **13**, 131-138

36. Bunt, A.H., Haschke, R.H., Lund, R.D. & Calkins, D.F. (1976) Factors affecting retrograde axonal transport of horseradish peroxidase in the visual system. *Brain Res.* **102**, 152-155

37. Oliver, J.M., Berlin, R.D. & Davis, B.H. (1984) Use of horseradish peroxidase and fluorescent dextrans to study fluid pinocytosis in leukocytes. *Methods. Enzymol.* **108**, 336-347

38. Renz, M. & Kurz, C. (1984) A colorimetric method for DNA hybridization *Nucleic. Acids Res.* **12**, 3435-3444

39. Durrant, I. (1990) Light-based detection of biomolecules *Nature (London)* **346**, 297-298

40. Stoecker, P.W. & Yacynych, A.M. (1990) Chemically modified electrodes as biosensors. *Selective Electrode Rev.* **12**, 137-160

41. Kulys, J. & Schmid, R.D. (1991) Bienzyme sensors based on chemically modified electrodes. *Biosensors Bioelectron.* **6**, 43-48

42. Misra, H.P. & Squatrito, P.M. (1982) The role of superoxide anion in peroxidase-catalysed chemiluminescence of luminol. *Arch. Biochem. Biophys.* **215**, 59-65

43. Meunier, G. & Meunier, B. (1985) Peroxidase-catalysed O-Demethylation reactions. *J. Biol. Chem.* **260** 10576-10582

44. Blum, L. (1993) Chemiluminescent flow injection analysis of glucose in drinks with a bienzyme fibreoptic biosensor. *Enzyme Microb. Technol.* **15**, 407-411

10

The renin-angiotensin system

Tadashi Inagami

Department of Biochemistry, Vanderbilt University School of Medicine, Nashville, TN 37232 U.S.A.

Introduction

Through many years of critical research the renin-angiotensin system (RAS) is now considered as the major regulator of blood pressure, electrolyte balance and renal, neuronal and endocrine functions related to cardiovascular control. Equally important is evidence that RAS is the key factor in a large majority of essential hypertension as indicated repeatedly by successes in treatment of hypertension by various inhibitors and receptor blockers of RAS. Renin was discovered in 1898 by Tegerstedt and Bergman as 'a pressor substance' produced in the kidney, with the connotation that it might be another hormone in the context of hormone research in the 19th Century. However, in 1940, Page & Helmer and Brann-Menendez *et al.* firmly established it as a circulating enzyme capable of producing a peptide hormone angiotensin (Ang) from a macromolecular prohormone angiotensinogen which is produced mainly in the liver. Later, Skeggs and his collaborators found an intermediate step involving inactive decapeptide Ang I, which is converted to the active octapeptide Ang II by an additional enzyme angiotensin I converting enzyme (ACE) (Figure 1).

Compared with other peptide hormone generating systems, RAS is unusual in that both the prohormone and specific cleavage enzyme are secreted into blood and the process of hormone formation (prohormone processing) takes place in plasma, whereas many hormones are produced in endocrine cells of their origin, However, more recent studies revealed that an intracellular mechanism of Ang formation also exists.

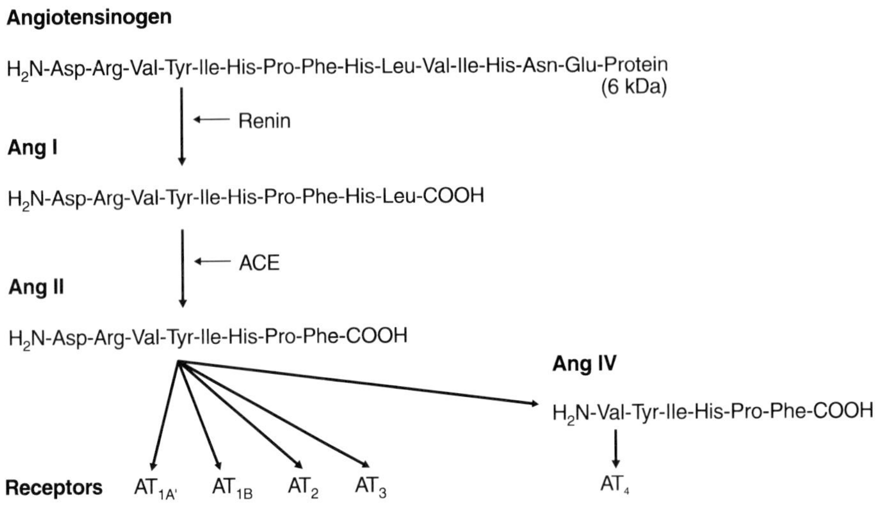

Angiotensinogen

H₂N-Asp-Arg-Val-Tyr-Ile-His-Pro-Phe-His-Leu-Val-Ile-His-Asn-Glu-Protein
(6 kDa)

Ang I

H₂N-Asp-Arg-Val-Tyr-Ile-His-Pro-Phe-His-Leu-COOH

← Renin

Ang II

← ACE

H₂N-Asp-Arg-Val-Tyr-Ile-His-Pro-Phe-COOH

Ang IV

H₂N-Val-Tyr-Ile-His-Pro-Phe-COOH

Receptors AT$_{1A'}$ AT$_{1B}$ AT$_2$ AT$_3$ AT$_4$

Figure 1. Pathways for the formation of active angiotensins from their prohormone angiotensinogen

Angiotensinogen

Angiotensinogen is a 56-60 kDa glycoprotein produced mainly in the liver and released into the blood[1]. It is the only known prohormone of Ang and is the only known substrate for renin. The Ang structure is localized in the *N*-terminus of this prohormone.

cDNA for rat angiotensinogen consists of a coding region of 1455 nucleotides encoding for 485 amino acid residues. The Ang I sequence immediately following a signal sequence was in agreement with the biochemical determination of amino acid sequences determined with purified human angiotensinogen. Human angiotensinogen genes encompass five exons of which exons II, III, IV, and V contain coding regions.

Angiotensinogen synthesis in the liver is regulated by slow and relatively moderate mechanisms. Glucocorticoids, oestrogen, thyroid hormone and Ang II stimulate the secretion of angiotensinogen. These hormones cause parallel changes in angiotensinogen mRNA and protein levels in the liver. Bilateral nephrectomy also suppresses angiotensinogen mRNA. The absence of angiotensinogen storage or secretory granules and cytosolic staining by antibodies to angiotensinogen suggested that it is released by a constitutive pathway rather than by a regulated secretory mechanism.

Although the primary organ of angiotensinogen synthesis is the liver, angiotensinogen mRNA is distributed widely in a variety of tissues which include vascular wall, renal proximal tubules, neuronal and astroglial cells. Angiotensinogen is found not only in plasma, but also in cerebrospinal fluid.

The vascular bed in rat hind-legs perfused with Krebs-Ringer buffer releases a large amount of angiotensinogen, indicating that various tissues can supply a relatively large amount of angiotensinogen for local production of Ang and even for the intracellular production of Ang as discussed later.

The plasma concentration of angiotensinogen is at a micromolar range which is not a saturating concentration for the enzyme renin. Thus, changes in angiotensinogen concentration can affect the rate of Ang production. Recent genetic analysis of siblings affected by essential hypertension indicated that the angiotensinogen gene difference is significantly correlated to human essential hypertension[2].

Renin

Renin has two unique features which have made it an exceedingly interesting enzyme for biochemical and physiological studies[3]. Its highly stringent substrate specificity limited to a singular leucyl peptide bond in angiotensinogen is exceptional. Furthermore, the mechanisms in the regulation of its releases from the juxtaglomerular (JG) cells of the kidney, the major source for plasma renin, are very tightly controlled. These processes are among the most stringent known for endocrine regulation. The mechanisms regulating renin release are all directed toward maintaining homoeostasis of blood pressure, electrolytes and blood volume. The stringent specificity allows renin to work on angiotensinogen selectively in plasma which contains numerous precursors for enzymes and hormones. Enzymologists had been baffled by the failure of type-specific protease inhibitors to inhibit renin, which was taken to indicate that it was a unique peptidase.

Numerous attempts at the purification of this interesting enzyme had been unsuccessful owing to its very low concentration and instability, until it was finally purified from the submandibular gland of male mice. Complete purification of renal renin, a more difficult task, required affinity chromatographic steps and the use of protease inhibitor cocktails for stabilization of renin against proteolytic loss[4]. The affinity chromatography of renin became feasible owing to the timely discovery of its interaction with pepstatin, a hexapeptide inhibitor of pepsin, and also by the use of haemoglobin Sepharose to minimize co-purification of renin and cathepsin D. A more specific inhibitor for renin, H-77 synthesized by Szelke, later simplified the purification work enormously in the laboratory of Leckie. Stable, pure renin preparations from pig, rat, dog and human kidney were obtained after greater than 100 000-fold purification.

While renin was refractory to inhibitors against serine-, cysteine- and metalloproteases, we found that it was completely inactivated by the specific aspartyl protease inhibitors diazoacetyl-D,L-norleucine methyl ester (DAN) in the presence of Cu^{2+} and 1,2-epoxy-3-p-nitrophenoxypropane (EPNP)[5]. In pepsin, these reagents had been shown to esterify the β-carboxyl groups of two different catalytically essential aspartyl residues in the active site independently. Although renin was not inhibited by pepstatin as strongly as pepsin, its susceptibility to DAN and EPNP clearly indicated that it was an aspartyl protease. We and others further confirmed this conclusion later by the determination of the complete amino acid sequences either by the direct Edman degradation method[6] or by deduction from the base sequence of its cDNA[7,8]. Human and mouse renin showed 41 and 43% sequence identity with porcine pepsin, respectively. Furthermore, the amino acid sequences in the vicinity of the two catalytically essential aspartyl residues were highly conserved among aspartyl proteases (Figure 2), whereas the overall amino acid sequences of renin and pepsin were only 40-45% identical.

The mouse genome in some strains contains two closely related renin genes (*Ren-1* and *Ren-2*), whereas other species have only one gene (*Ren-1*). *Ren-1* is expressed in the JG cells of the kidney throughout the entire mammalian species. The products of the *Ren-1* and *Ren-2* genes are very similar in their structures.

Figure 2. High degree of amino acid sequence identity in two catalytically essential aspartyl (acid) protease
Asp residues indicated in red in the top panel represent one of the catalytically essential Asp residues which are esterified by DAN methylester. It corresponds to Asp^{215} of pepsin. Asp in the lower panel is specifically inactivated by EPNP which corresponds to Asp^{32} of pepsin.

```
                                          DAN
                                           ↓
Human renin          - Leu - Val - Asp - Thr - Gly - Ala - Ser -

Mouse renin          - Val - Val - Asp - Thr - Gly - Ser - Ser -

Rat renin            - Val - Val - Asp - Thr - Gly - Thr - Ser -

Human cathepsin D    - Ile - Val - Asp - Thr - Gly - Thr - Ser -

Pig pepsin           - Ile - Val - Asp - Thr - Gly - Thr - Ser -

Bovine chymosin      - Ile - Leu - Asp - Thr - Gly - Thr - Thr -

Penicillopepsin      - Ile - Ile - Asp - Thr - Gly - Thr - Thr -


                                         EPNP
                                           ↓
Human renin          - Val - Phe - Asp - Thr - Gly - Ser -

Mouse renin          - Ile - Phe - Asp - Thr - Gly - Thr -

Rat renin            - Ile - Phe - Asp - Thr - Gly - Ser -

Human cathepsin D    - Val - Phe - Asp - Thr - Gly - Ser -

Pig pepsin           - Ile - Phe - Asp - Thr - Gly - Ser -

Bovine chymosin      - Leu - Phe - Asp - Thr - Gly - Ser -

Penicillopepsin      - Asn - Phe - Asp - Thr - Gly - Ser -
```

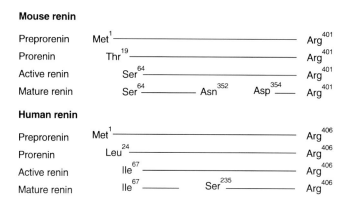

Figure 3. Processing of preprorenin to mature renin

As shown in Figure 3, preprorenin synthesized from human *Ren-1* and mouse *Ren-2* mRNA contains 406 and 401 amino acid residues, respectively, which are converted to prorenin with 383 amino acid residues. It is further converted to active forms by the cleavage of pro-sequences (43 and 45 residues, respectively). Active renin is further processed proteolytically to end-products which have a cut in the middle (human renin) or in the N-terminal region (rat and mouse renin). The enzyme responsible for the prorenin activation is being pursued. Several different types of proteases have been proposed as candidates.

Model-building studies of renin, based on the amino acid sequence homology of mouse or human renin with fungal aspartyl proteases, and X-ray crystallographic studies of human and rat renin, without their carbohydrate moieties produced by genetic engineering, revealed that the active enzyme consists of a bilobal structure, each of the lobes containing one of the two catalytically essential aspartyl residues in juxtaposition to each other in the major cleft formed between the two lobes (Figure 4)[9]. Each of the lobes is composed largely of β-pleated sheet structure and the lobes are connected to each other by a single polypeptide bridge. The amino acid sequence of the two lobes indicates a degree of similarity, suggesting possible gene duplication. The presence of genes encoding only one-half of the aspartyl proteases in retroviral genomes indicates that the two lobes can associate to form an active dimeric enzyme. In addition to its exceedingly stringent substrate specificity, it is interesting that this aspartyl protease is active in a neutral pH range in contrast to other aspartyl (acid) proteases, such as pepsin, chymosin and cathepsin D, which are all active in acidic pH ranges. No specific structural features have yet been identified to explain this property.

The long cleft formed by the two lobes will provide subsites for recognizing and binding a long N-terminal sequence of angiotensinogen spanning seven to eight amino acid residues. It is believed that such an extensive subsite structure is responsible for the unusually stringent substrate

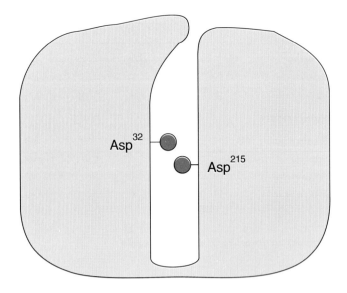

Figure 4. Bilobal structure of renin with a deep and long groove
Two catalytically essential carboxyl groups of aspartyl residues are in juxtaposition to each other.
The groove is covered by a large flap indicated by the shaded area.

specificity of renin. On the other hand, it is believed that the long inhibitory
N-terminal segment of prorenin (the prosegment consisting of more than 40
amino acid residues) may wrap and/or insert itself in the region of this cleft
before the proteolytic activation of prorenin.

   Prorenin was discovered as proteolytically activatable inactive renin in
amniotic fluid and plasma. Purification of inactive renin from plasma, kidney
and human chorion membrane, followed by determination of the N-terminal
sequence and its activation by proteolysis, produced unequivocal evidence that
this inactive renin is a prorenin. This conclusion was supported by the base
sequences of preprorenin cDNA[10].

Control of renin activity

Since renin is at the head of the renin-angiotensin cascade and its quantity is
several orders of magnitude lower than those of its substrate angiotensinogen
or ACE, which catalyses the subsequent step of the Ang II formation, renin is
considered to be the rate-limiting step in the production of Ang. However,
active renin is a simple enzyme without any allosteric control mechanism.
Although conversion of prorenin to active renin provides a means of
regulation of its enzyme activity, pathways for the secretion of prorenin and
active renin are distinct. Prorenin is secreted by a constitutive pathway,
whereas active renin is secreted by a regulated mechanism. Other than the
possible participation of a renin-binding protein or renin inhibitor, this

regulated secretion seems to be the dominant mechanism in controlling the renin activity in plasma.

Stable and sufficient blood flow is essential for the maintenance of the life process. It requires maintenance of not only blood pressure, but also of blood volume, which is regulated by the homoeostasis of electrolyte (mainly Na^+ and Cl^-) balance.

The nervous system plays an important role in the homoeostasis. Renin release from its major site of production and storage, the JG cells of the afferent arteriole of the kidney, is stimulated by the β_2-receptor-mediated component of the renal adrenergic activity.

The renal arteries, arteriole or perhaps even the JG cells can sense the blood pressure directly by a putative local baroreceptor. Reduced blood pressure is sensed by the endothelium and seems to activate prostacycline formation. The prostacycline readily reaches the JG cells and activates adenylyl cyclase through the prostacycline receptor. Cyclic AMP thus produced in JG cells elicits the release of renin from JG cells. Thus, both the β_2-adrenergic and prostacycline-mediated stimulation of renin release are mediated by this cyclic AMP mechanism. The baroreceptor can also change the membrane potential of JG cells, which in turn modulates renin release.

Increased salt intake reduces plasma renin markedly through increased Cl^- concentration in the renal tubular fluid. Since Ang II activates Na^+ re-absorption from the renal tubules, inhibition of renin release, as the result of the increased NaCl in the renal tubular fluid, is important in the maintenance of the blood volume homoeostasis. The anatomy of distal convoluted tubules is such that it is in contact with JG cells in the macula densa region of the polar tufts of renal glomerulus. It is believed that a Cl^--sensitive change in membrane potential of JG cells is involved in the inhibition of renin release. Ang II contributes to Na^+ retention significantly by way of renal action of aldosterone, which is released from the adrenals by the action of Ang II. Ang II directly acts on JG cells and inhibits renin release partially through calcium-calmodulin-dependent activation of nitric oxide synthesis. These mechanisms represent one of the most exquisitely regulated mechanisms of endocrine regulation[11].

Sites of renin action

Although renin released from the kidney was thought to catalyse Ang I production in plasma, recent studies on the rate of Ang production in the vasculature of various organs indicate that the majority of Ang I production is owing to renin associated with vascular cells and tissues. The association of renin to vascular tissue has been reported by Swales and his associates[12]. Our preliminary observation indicates renin-binding capacity on the surface of vascular smooth muscle cells in culture. It is likely that renin is associated to these sites, and produces Ang I, allowing an immediate access to the vascular

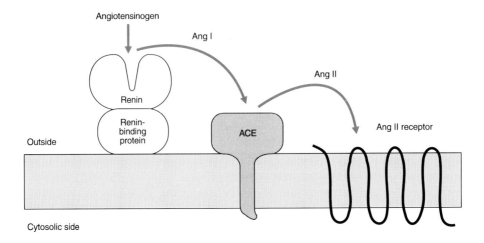

Figure 5. Schematic representation of a Ang II-forming process mediated by renin and ACE bound to the outer surface of vascular cells
Transfer of Ang I and II on the surface of the membrane proceeds much more efficiently than free collision in solution. The presence of renin-binding proteins has been supported by several observations.

cell-associated ectoenzyme ACE which, in turn, permits very efficient production of Ang II in the immediate vicinity of its receptor on the surface of plasma membrane of vascular smooth muscle and endothelial cells. As shown in Figure 5, the kidney-derived plasma enzyme like renin may perform most of its catalytic function on the surface of vascular tissues. Two-dimensional migration of molecules in solid-phase catalyst surface is known to be orders of magnitude faster than free collisions in solution.

In addition to the mechanism of extracellular Ang II production on the surface of vascular cells, intracellular mechanisms of Ang II production have been demonstrated in various endocrine and neuronal tissues. The presence of Ang II in the brain neurons, pituitary-luteinizing-hormone-producing cells, testicular Leydig cells and ovarian thecal cells are but a few examples. The use of cultured cells as models has shown that these cells and tissues seem to have the ability to produce renin, angiotensinogen and ACE. Angiotensinogen can be taken up also from interstitial fluid as it is produced by other cells. The concentrations of renin and its mRNA are independent of plasma (e.g. as shown for renin in the adrenal and brain), strongly indicating its local production rather than an origin from the kidney.

The presence of Ang II at very high concentrations in renal proximal tubular fluid and epididymal tubular fluid (in which rapid fluid transport takes place) is particularly interesting. Local production of Ang II seems to be the most reasonable explanation for the locally elevated Ang II concentration, which is much higher than that in plasma.

To determine possible roles of extra-renal tissue renin, Mullins *et al.* produced transgenic rats with the mouse *Ren-2* gene incorporated into their genome. Since the *Ren-2* gene is under a distinct regulatory mechanism from the rat *Ren-1* gene, the *Ren-2* gene may have been expressed in certain strains at an abnormal level. This produced a strain with very high 'fulminant' blood pressure levels. It is noteworthy that in these animals, production of its native renal renin was minimal. Instead, some strains exhibit highly elevated renin in their adrenal cortex. Transgenic rats with the human renin gene or human angiotensinogen gene alone were not hypertensive. This is because rat angiotensinogen and renin do not react well with human renin or angiotensinogen, respectively. However, when both were expressed in an animal, they also developed hypertension without any increase in the renal renin expression. This, again, suggests that renin and angiotensinogen expressed in tissues other than the kidney can contribute to the presence of functionally active Ang II at strategic sites, which results in hypertension by activating angiotensin receptors in, as yet, unidentified areas. These observations present strong evidence for the importance of locally generated Ang II either by intracellular renin or by renin (probably derived from the kidney) associated with the external surface of cells in strategic tissues. However, this is an artificial system resulting from abnormal expression of the renin gene. For identifying this mechanism with the underlying cause in essential hypertension, the burden of proof is on us to demonstrate that simultaneous over-expression of the genes of renin and angiotensinogen occurs in certain cells in essential hypertensives or animal models of hypertension.

ACE

This enzyme was discovered by Skeggs and his co-workers during their studies to isolate Ang. Helmer showed that both ACE and the bradykinin-inactivating enzyme kininase II are inhibited by snake venom peptides, the bradykinin-potentiating factors, and suggested that these two enzymes are identical or closely related. Concrete evidence for their identity was obtained after the complete purification of these enzymes[14].

This discovery revealed a very important conceptual significance to the regulatory function of ACE/kininase II as it connects the production of vasoconstrictor Ang II and inactivation of vasodilator bradykinin. Thus, ACE action shifts the balance toward vasoconstriction (Figure 6). This aspect becomes more apparent as the use of ACE inhibitors as effective therapeutic agents finds a wider application not only for renovascular hypertensive patients, or hypertensives with high plasma renin activity, but also for a much wider range of essential hypertensive patients, as will be discussed later. Further studies of ACE revealed that it is a Zn^{2+}-requiring metallopeptidase which also requires Cl^- ion for its full activity. Although its activity may be described as dipeptidyl carboxypeptidase, and its action stops at the prolyl residues in the *C*-terminal region of Ang I (- His-Pro-Phe-His-Leu-COOH), later studies of its structure indicated that it belongs to the family of neutral

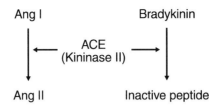

Figure 6. Dual significance of ACE in its pressor effects

metallo-endopeptidases. Although ACE was discovered as a plasma enzyme, its localization in the vascular bed, particularly in the lung by Vane and his associates, revealed that it is a membrane-bound ectoenzyme mainly localized on the surface of vascular endothelial cells.

Since ACE is a membrane-bound ectoenzyme, its purification required solubilization by detergents or by treatment with trypsin. It is a glycoprotein consisting of a single polypeptide chain. The relative molecular mass of human ACE ranges from 140 000 to170 000, presumably reflecting a difference in the size of the carbohydrate moiety. The amino acid sequence of ACE deduced from the base sequence of cDNA for human ACE revealed a very interesting feature. It has a relative molecular mass of 146 600. As shown in Figure 7, the major portion on the *N*-terminal side of the molecule is projected into the extracellular space. It is followed by a short, membrane-spanning domain of 17 amino acid residues, then by a short cytoplasmic tail of 30 amino acid residues in the *C*-terminus[15].

Interestingly, the extracellular portion contains two active-site domains, each containing the canonical sequence of His-Glu-Met-Gly-His, which chelates a Zn^{2+} ion and forms a catalytic site. Although the *C*-terminal domain was earlier thought to be the only active domain, both domains seem to be active. The *C*-terminal domain seems to have a greater $k_{cat.}$ value than that of the *N*-terminal domain. The activity of the *N*-terminal domain, however, seems to be less dependent on Cl^-.

cDNA cloned from the testis encodes for a smaller ACE molecule. Southern blot analysis of the ACE gene indicated that ACE has only one gene. Studies on the exon-intron structure of the ACE gene revealed that the cDNA for the testicular ACE is a product of alternative splicing of 26 exons. All but exon 13 is transcribed in the mRNA for the somatic form, whereas in the germinal cells, ACE mRNA starts from exon 13 to exon 26. Thus, the testicular ACE contains only the *C*-terminal active site. The physiological significance of the truncation in the testicular ACE is not clear[15]. In human heart the conversion seems to be mediated by a chymase-like enzyme.

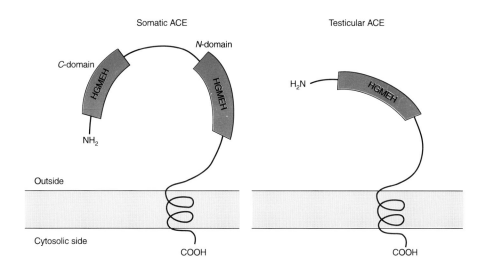

Figure 7. Structures of ACE in somatic and testicular cells
The amino acid sequence HGMEH (His-Gly-Met-Glu-His) represents the Zn^{2+}-binding sites in the catalytic site of ACE. Constructed on the basis of data in reference 15.

ACE inhibitor and pathophysiology of the RAS

The structural requirement for a substrate of ACE is relatively straightforward. Peptides with a free C-terminal carboxylate anion (not an absolute requirement) serve as the substrates. The penultimate peptide bond is selectively cleaved. By drawing analogy from pancreatic carboxypeptidase A, the C-terminal carboxyl group was considered to be anchored to a positively charged basic group and the scissile peptide bond must be placed in juxtaposition to the Zn^{2+} ion in the active site. An inhibitor to fit into the space was designed as shown in Figure 8. The thiol group would have a strong affinity for the heavy metal ion Zn^{2+}, whereas the proline residue in the C-terminal was expected to confer resistance against proteolytic degradation. This simple inhibitor, named captopril, with a dissociation constant of 23 nM for ACE, turned out to be one of a few successful drugs specifically designed on the basis of the specific structural requirement for an enzyme inhibitor/substrate[16].

This inhibitor of ACE was originally designed to inhibit the production of Ang II for the treatment of renovascular hypertension in which the plasma renin level is unusually elevated as a result of stenosis in the renal artery. This causes a reduction in renal blood flow and pressure, thereby activating the baroreceptor-mediated release of renin. It soon became clear that captopril was effective in normalizing the blood pressure of essential hypertensives with normal or even low plasma renin activity. Spontaneously hypertensive rats (SHR) and their stroke-prone relatives (SHRSP) bred by Okamoto, Aoki and Yamori, which are considered as models of human essential hypertension (the

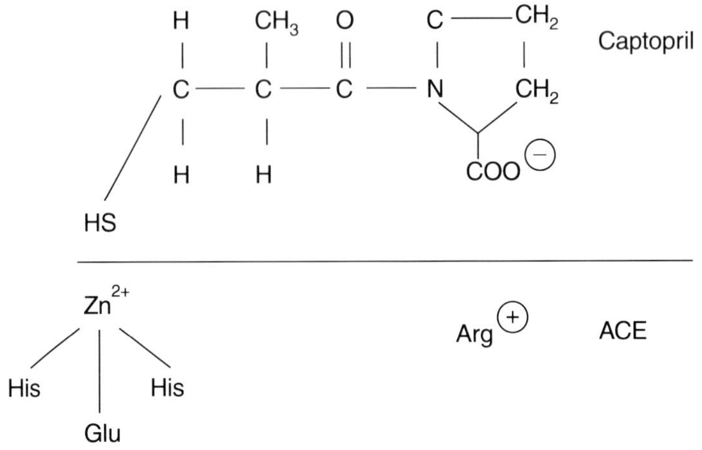

Figure 8. Schematic representation of the active site structure of angiotensin converting enzyme and its inhibitor captopril

plasma renin level of SHR is subnormal), also responded to captopril and their blood pressure was normalized.

These observations raised very important questions concerning the role of renin-angiotensin in essential hypertension. It indicated that high plasma renin activity is not an indicator for the aetiological role of renin-angiotensin in hypertension. Is it then renin, in Ang II-sensitive tissues, such as the hypothalamus, adrenal or blood vessel wall, that causes hypertension? Renin has been shown to perform its catalytic function both inside cells or on the surface of cells to which it can be adsorbed. Increased sensitivity to Ang II, either by elevated receptor responsiveness, increased receptor concentration, hypertrophied blood vessels, or by neuronal mechanisms, could also explain the hypertensive mechanism of angiotensin. Whatever the mechanisms may be, the importance of the RAS over other pressor systems is recognized. Effective blood pressure normalization is not limited to ACE inhibitors. Inhibitors of renin and blockers of Ang II receptors were equally effective in the treatment of many, but not all, essential hypertensives. They were also effective in normalizing the blood pressure of SHR. This has not only opened a Pandora's box in hypertension research, but it has also shown that the RAS is a gold mine for developing antihypertensive drugs.

Angiotensin

Ang II structure was determined in 1954 by Peart *et al.* and Skeggs *et al.* and Ang I by Skeggs and his associates (Figure 1). Subsequent complete synthesis of Ang II by Bumpus *et al.* and Schwyzer *et al.* made significant contributions to the delineation of the action of Ang II by making available a sufficient supply of the peptides for pharmacological studies. Of equal significance was the synthesis of inhibitory analogues of Ang II such as [Sar^1, Ala^8]-Ang II

(Saralasin), and [Sar^1, Ile^8]-Ang II (Sarile) and subsequent non-peptide inhibitors, which turned out to be specific for isoforms of the Ang II receptor as discussed later.

Actions of Ang II and Ang II receptors

Ang II was first recognized for its potent vasoconstriction *in vivo* and *in vitro*. Another very important action of the peptide on the stimulation of release of aldosterone from the adrenals was reported in 1960-61. Since aldosterone causes sodium retention by the kidney, the dual action of Ang II (vasoconstriction and salt retention) was considered as the unique feature of Ang II in initiating and maintaining hypertension. Further studies revealed the diverse responses that Ang II elicits in various cells and tissues[17]. Although a detailed and exhaustive description of these mechanisms is clearly beyond the scope of this essay, some of the angiotensin-elicited responses are listed in Table 1.

Isolated and cultured cells permitted studies on the detailed biochemical mechanisms of these responses. It has become clear that most of the responses can be explained by increased cytosolic Ca^{2+} released from either intracellular stores by transiently increased inositol trisphosphate owing to the action of phospholipase C (a process mediated by G_q-protein), or by influx of the extracellular Ca^{2+} through various types of Ca^{2+} channels (mediated by G_o-

---

Table 1. Responses elicited by Ang II

- Contraction of vascular and other smooth muscles and inotropic effects on cardiac myocytes (activation of phospholipase $C\beta_1$ and opening of L-type Ca-channels)
- Stimulation of release of various hormones, such as mineral and glucocorticoids from adrenal cortex, catecholamines from adrenal medulla, vasopressin and ACTH from the pituitary
- Inhibition of renin release from JG cells of the kidney
- Facilitation of noradrenaline release from peripheral nerve endings
- Stimulation of endothelial cells and other tissues for the production of prostaglandins and other arachidonate metabolites (activation of phospholipase A_2)
- Direct effect on renal tubular cells to stimulate Na re-absorption or to reduce it at different Ang II concentrations (including activation of Na-H antiporter)
- Stimulation of hypertrophy of vascular and cardiac myocytes (and sometimes mitogenesis of vascular and fibroblast cells)
- Brain-mediated water drinking
- Centrally mediated hypertension owing to increased adrenergic out-flow
- Diverse effects on reproductive systems including uterus, placenta, testis and ovary by, as yet, poorly defined mechanisms
- Development of fetus
- Glycogenolysis in heptatocytes

protein). In certain cells, a decrease in cyclic AMP production has been noted (mediated by G_i-protein). A chain of events follows these initial responses, i.e. stimulation of protein kinase C, phospholipases A_2 and D, and initiation of mitogenesis and/or protein synthesis (cellular hypertrophy). All of these receptor functions are mediated by the Ang II receptor subtype 1 (AT_1) receptor isoform through its interaction with G_q, G_i or G_o.

The presence of specific Ang II receptors with high affinity (nM K_D values) for Ang II was demonstrated in plasma membranes of hepatocytes, vascular smooth muscles, and adrenal zona glomerulosa cells (in the outermost layer of the cortex where aldosterone is synthesized). Synthesis of peptidic and non-peptidic Ang II receptor antagonists in 1989 and pharmacological studies revealed that there are two pharmacologically distinct receptor isoforms[18]. Owing to their exceeding instability, their study required the expression cloning method.

The cDNA for rat and bovine AT_1[19,20] was cloned. Two closely related AT_1 subtypes, AT_{1A} and AT_{1B} have been reported. While their coding regions are 95% similar, the expression of AT_{1B} seems to be regulated by a mechanism different from that for AT_{1A}, as the structures of their promotor regions are quite different. Functionally, AT_{1B} does not interact with G_i protein. As expected from its G-protein-coupled functions, it showed the structural motif for a typical seven-transmembrane domain receptor. It showed binding specificity typical for AT_1 receptor preferentially binding to the AT_1-specific non-peptidic antagonist losartan, but not to the AT_2 isoform specific antagonists (CGP 24112A and the PD series). When transiently expressed in COS-7 cells, Ang II elicited transient production of inositol trisphosphate and Ca^{2+} release from internal storage followed by a sustained phase of Ca^{2+} elevation — presumably through G_o-protein-mediated opening of a hormone-operated Ca^{2+} channel. In addition, the AT_{1A} receptor permanently expressed in Chinese hamster ovary cells revealed evidence for coupling to G_i-proteins as evidenced by inhibition of adenylyl cyclase. Thus, a single AT_{1A} receptor can interact with three different G-protein α-subunits. The mechanisms involved in such multiple interactions are certainly worth investigating. Ang II binding sites are being clarified; this small peptide seems to find its way into a space created by a well-like structure formed by the putative seven helical columns(Figure 9).

In addition to AT_1 isoforms, there are three clearly distinct receptor isoforms, AT_2, AT_3 and AT_4 isoforms that have been identified by a pharmacological approach. As summarized in Table 2, the AT_2 isoform shows specific binding to CGP 42112A and the PD series of receptor antagonists. Among the so-called AT_2-specific antagonist, CGP 42112A seems to be an agonist rather than antagonist. This receptor is expressed at high levels in the mesenchymal tissues of the fetus, uterus, rat adrenal medulla and certain specific brain regions such as locus coeruleus and inferior olive. The expression of this receptor is closely associated with growth and/or differentiation, since

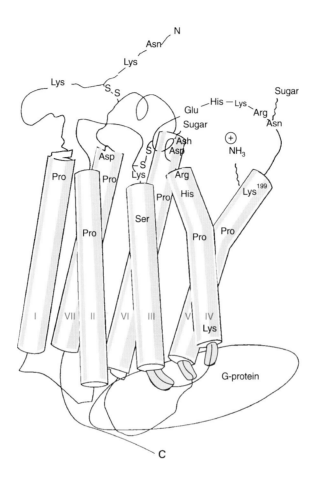

Figure 9. A well-like structure of AT₁ receptor by computer graphics
Note Lysin[199] is an anchoring site for the C-terminal carboxylate anion.

AT₂ concentration increases markedly in cultured vascular smooth muscle cells and endothelial cells under quiescent conditions (serum-free medium) and in fetus in the mid-gestation period. It inhibits some phosphotyrosine phosphatases.

AT₃ and AT₄ are defined for their lack of affinity to both AT₁ and AT₂ specific antagonists. The agonist for AT₃ is angiotensin II, whereas the agonist for AT₄ is the hexapeptide angiotensin IV (Val-Tyr-Ile-His-Pro-Phe). AT₃ was found in neuroblastoma cells, whereas AT₄ is expressed extensively in vascular endothelium, vascular smooth muscle, and brain cortex regions. In both AT₃ and AT₄, the agonists seem to activate nitric oxide (NO) synthesis. NO is a well-known smooth muscle relaxant and inhibitor of mitogenesis through the stimulation of cyclic GMP synthesis. Thus their physiological significance seems to be opposite to the vasoconstrictor, hypertrophic and mitogenic action of AT₁. The balance in the expression of these receptors seems to determine the net effects of Ang II in various tissues. Since different tissues contain mixed

Table 2. Ang II receptor isoforms

	Inhibitors			Coupled to:	Features
	Losartan	PD123177	CGP 42112		
AT_{1A} & AT_{1B}	Yes	No	No	G_q, G_i, G_o	Seven-transmembrane structure
AT_2	No	Yes	Agonist	G-proteins	Seven-transmembrane structure
AT_3	No	No	No	—	Unknown
AT_4	No	No	No	—	Unknown, Ang IV hexapeptide C is agonist

population of these receptors, the evaluation of the effects and intracellular signalling mechanism mediated by these receptors requires a pure population of each receptor. Thus cloning of each isoform of the receptor is essential for further development of our studies on Ang II receptors.

The diversity and complexity of responses to Ang II were discussed above. The versatility of receptors may provide a partial explanation for this diversity of function. In addition, highly sensitive desensitization (tachyphylaxis) and down-regulation are important for the regulation of Ang II actions. Complex regulatory mechanisms of the receptor function and expression are being revealed as the result of many years of studies on Ang II receptors in various tissues, and from cloning and mutagenesis studies. Most of them are negative regulatory mechanisms and seem to involve at least four separate mechanisms as shown in Table 3.

The first three mechanisms may involve some specific changes in the receptor proteins owing to binding of specific agonist, but the detailed mechanisms involved remain to be clarified. The down-regulation of gene expression should be owing to the regulation of the promotor function of Ang receptor genes by Ang II-responsive and other mechanisms. Various positive enhancer regions (e.g. glucocorticoid responsive elements) as well as negative elements (AP-1) are being identified.

Table 3. Mechanism of regulation of Ang II receptor AT_1

1.   Rapid uncoupling from G-protein to shift the receptor to a low-affinity form

2.   Sustained deprivation of the G-protein coupling ability (desensitization or tachyphylaxis occurring in a few minutes)

3.   Internalization of the receptor-agonist complex and possible turnover of the receptor (10-30 min)

4.   Down-regulation of receptor gene expression (1-2 h)

Summary

Unravelling of the molecular mechanisms of the action of RAS has been slow. Nature has been rather stingy in revealing bits and pieces of information. Each step of development has depended on the innovation of an appropriate methodology. The uniqueness of the RAS lies in:

- *The function and regulation of the highly specific enzyme renin which specifically catalyses the conversion of the prohormone angiotensinogen to Ang I by an extracellular mechanism.*
- *The production of the agonist Ang II takes place in two steps.*
- *Ang II and its metabolites exert exceedingly diverse pathophysiological effects, presumably through the complex and multifunctional receptors.*

The exquisite mechanisms involved in the regulation of renin release and receptor regulation are fascinating. The intricate mechanisms that nature has devised for the checks and balances to maintain steady blood flow and electrolyte balance present a great challenge to biochemists in their attempts to clarify the mechanisms involved at both molecular and cellular levels.

In relation to the pathophysiology of hypertension, particularly essential hypertension, there is no question that the RAS plays a pivotal role. Although numerous mechanisms could explain its hypertensinogenic effects, no single mechanism can be identified as the major determinant at the present stage of our knowledge. However, there is an important consensus that the effect of Ang II is manifested slowly at even subpressor doses of Ang II through long-term effects involving remodelling of the cardiovascular and renal system.

References

1. Menard, J., Clauser, E., Bounik, J. & Corvol, P. (1993) Angiotensinogen: biochemistry, in *The Renin Angiotensin System* (Robertson, J.I.S. & Nichols, M.G., eds.), pp. 8.1-8.10, Gower Medical Publishing, London

2. Jeunemaitre, X., Soubrier, F., Kotelevisev, X.V., Lifton. R. P., Williams, C.S., Charru, A., Hunt, S.C., Hopkins, P.N., Williams, R. R., Lalonel, J. M. & P. Corvol (1993) Molecular basis of human hypertension; role of angiotensinogen. *Cell* **71**, 169-180

3. Inagami, T. (1993) Renin: purification, structure and function, in *The Renin Angiotensin System* (Robertson, J.I.S. & Nicholls, M.G., eds.), pp. 4.1-4.17, Gower Medical Publishing, London

4. Murakami, K. & Inagami, T. (1975) Isolation of pure and stable renin from hog kidney. *Biochem. Biophys. Res. Commun.* **62**, 757-763

5. Inagami, T., Misono, K. & Michelakis, A.M. (1974) Definitive evidence for similarity in the active site of renin and acid protease. *Biochem. Biophys. Res. Commun.* **56**, 503-509

6. Misono, K.S., Chang, J.J. & Inagami, T. (1982) Amino acid sequence of mouse submaxillary gland renin. *Proc. Natl. Acad. Sci. U.S.A.* **79**, 4858-4862

7. Panthier, J.J., Foote, B., Chambrand, B., Strossberg, A.D., Corvol, P. & Rougeon, F. (1982) Complete amino acid sequence and maturation of the mouse submaxillary gland renin precursor. *Nature (London)* **298**, 90-92

8.   Imai, T., Miyazaki, H., Hirose, S., Hori, H., Hayashi, T., Kageyama, R., Ohkubo, H., Nakanishi, S. & Murakami, K. (1983) Cloning and Sequence analysis of cDNA for human renin precursor. *Proc. Natl. Acad. Sci. U.S.A.* **80**, 7405-7409

9.   Sielecki, A.R., Hayakawa, K., Fujinaga, M., Murphy, M.E.P., Fraser, M., Muir, A.K., Carilli, C.T., Lewicki, J.A., Baxter, J.D. & James, M.N.G. (1989) Structure of recombinant human renin: a target of cardiovascular active drug at 2.5Å resolution. *Science* **243**, 1346-1351

10. Schalekamp, M.A.D.H. & Derkx, F.H.M. (1993) Biochemistry of prorenin, in *The Renin Angiotensin System* (Robertson, J.I.S. & Nicholls, M.G., eds.), pp. 6.1-6.13, Gower Medical Publishing, London

11. Katz, S.A. & Malvin, R.L. (1993) Renin secretion: control, pathways and glycosylation in *The Renin Angiotensin System* (Robertson, J.I.S. & Nicholls, M.G., eds.), pp. 24.1-24.13, Gower Medical Publishing, London

12. Loudon, M., Bing, R.T., Thurston, H. & Swales, J.D. (1983) Arterial wall uptake of renal renin and blood pressure control. *Hypertension* **5**, 629-634

13. Mullins, J.J., Peters, J. & Ganten, D. (1990) Fulminating hypertension in transgenic rat harboring mouse Ren-2 gene. *Nature (London)* **344**, 541-544

14. Yang, H.Y.T., Erdös, E.G. & Levin, Y. (1970) a dipeptidyl carboxypeptidase that converts angiotensin I and inactivates bradykinin. *Biochim. Biophys. Acta* **214**, 374-376

15. Soubrier, F., Alhenc-Gelas, F., Hubert, C., Allegrini, J., John, M., Tregear, G. & Corvol, P. (1988). Two putative active centers in human angiotensin I-converting enzyme revealed by molecular cloning. *Proc. Natl. Acad. Sci., U.S.A.* **85**, 9386-9390

16. Ondetti, M.A., Rubin, B. & Cushman, D. (1977) Design of specific inhibitors of angiotensin converting enzyme: new orally active antihypertensive agent. *Science* **196**, 441-444

17. Catt, K.J. (1993) Angiotensin II receptors, in *The Renin Angiotensin System* (Robertson, J.I.S. & Nicholls, M.G., eds.), pp. 12.1-12.14

18. Timmermans, P.B.M.W.M., Wong, P.C., Chiu, A.T. & Herblin, (1991) Nonpeptide angiotensin II receptor antagonist. *Trends Pharmacol. Sci.* **12**, 55-61

19. Sasaki, K., Yamano, Y., Bardhan, S., Iwai, N., Murray, J.J., Hasegawa, M., Matsuda, T. & Inagami, T. (1991) Cloning and expression of a complementary DNA encoding a bovine adrenal angiotensin II type-1 receptor. *Nature (London)* **351**, 230-233

20. Murphy, T.J., Alexander, W.R., Griending, K.K., Runge, M.S. & Bernstein, K.E. (1991) Isolation of a cDNA encoding the vascular type-1 angiotensin II receptor. *Nature (London)* **351**, 253-256.

Subject index